Accelerators and Colliders

Edited by Ozan Artun

Published in London, United Kingdom

IntechOpen

Supporting open minds since 2005

Accelerators and Colliders
http://dx.doi.org/10.5772/intechopen.87914
Edited by Ozan Artun

Contributors
Kazuhisa Nakajima, Min Chen, Zhengming Sheng, Nathalie Olivi-Tran, Leonard Kisslinger, Jie Wang, Sheng Wang, Ozan Artun

First published in London, United Kingdom, 2020 by IntechOpen
IntechOpen is the global imprint of INTECHOPEN LIMITED, registered in England and Wales, registration number: 11086078, 7th floor, 10 Lower Thames Street, London, EC3R 6AF, United Kingdom
Printed in Croatia

British Library Cataloguing-in-Publication Data
A catalogue record for this book is available from the British Library

Additional hard and PDF copies can be obtained from orders@intechopen.com

Accelerators and Colliders
Edited by Ozan Artun
p. cm.
Print ISBN 978-1-83968-262-9
Online ISBN 978-1-83968-263-6
eBook (PDF) ISBN 978-1-83968-264-3

For EU product safety concerns:
IN TECH d.o.o., Prolaz Marije Krucifikse Kozulić 3, 51000 Rijeka, Croatia, info@intechopen.com or visit our website at intechopen.com.

Meet the editor

Dr. Ozan Artun obtained both a bachelor's degree and a master's degree in Physics from Trakya University, Edirne, Turkey. He pursued a PhD in Physics at Yildiz Technical University, Istanbul. After one year, he was appointed to Bülent Ecevit University as a research assistant and received his PhD in Nuclear Physics in 2015. Dr. Artun also has a master's degree in Computer Engineering. In 2018, he became an associate professor. He is also a reviewer for many international journals. His research interests include particle accelerators, radioisotope production, nuclear batteries, software development, nuclear structure, nuclear fuel, and ionizing radiation.

Contents

Preface

Since the mid-twentieth century, accelerators and colliders have led the development of science and technology in fields of space, medicine, and energy.

Understanding the applications of existing accelerators and colliders with different energies is very important for understanding developments in fundamental or advanced science and technology, such as exploration of the subatomic world, interaction of particles, structure of matter, and medical and energetic radioisotope productions. This book takes into account the currently available types of accelerators and colliders, like linear and circular accelerators and colliders, and their many applications as have been presented in scientific publications. In addition, the book also discusses those accelerators and colliders currently under construction and investigation.

The first chapter provides a short introduction to accelerators and colliders across the world, while the second chapter presents applications for production of the medical radionuclide Ir-192, which is used in brachytherapy. The third chapter explains vacuum systems as essential parts of accelerators and colliders and reviews the key technologies involved in optimizing vacuum system designs with a focus on high-energy, high-intensity, and high-luminosity accelerators and colliders. The fourth chapter defines advanced collider systems and includes a novel scheme of a gas-filled capillary accelerator driven by a laser pulse formed from two-mode mixing of capillary eigenmodes and proposes a laser-plasma linear collider. The fifth chapter presents theoretical calculations of the masses of elementary fermions and a classification of elementary particles over all space and temporal dimensions, including the theoretical values of masses for all the elementary fermions (electrons, muons and taus; all quarks and all neutrinos). The last chapter is a review of elementary particles, Quantum Chromodynamics (QCD), and strong interactions in QCD theory via gluon exchange between quark–antiquark-producing mesons.

Each chapter in the book contributes sophisticated knowledge about accelerators and colliders and their crucial applications for current and forthcoming technology. I would like to thank the authors for their contributions. I am also grateful to IntechOpen Service Manager Ms. Lada Bozic for her patience and support.

Dr. Ozan Artun
Associate Professor,
Zonguldak Bülent Ecevit University,
Department of Physics,
Division of Nuclear Physics,
Zonguldak, Turkey

Chapter 1

Introductory Chapter: Accelerators and Colliders

Ozan Artun

1. Introduction

Accelerators are devices that may repel charged particles such as protons and electrons near the speed of light. The charged particles are collided either on targets or towards other particles in the opposite directions. Accelerators utilize electromagnetic fields to accelerate the charged particles, and radio-frequency cavities increase the particle beams as magnet in accelerators focus the beams and curves to their trajectory. There are significant properties of an accelerator according to the aim of usage, e.g., the energy of collisions and type of particles. Therefore, the number of accelerators in operation around the world exceed 30,000. It is obvious that the need of understanding the nature and determination of nature's laws in the subatomic dimension have been provided by accelerators especially in particle physics, because the developments in particle accelerators and particle detectors ensure attractive opportunities for great scientific advances [1, 2]. In addition to researches in physics, the particle accelerators are used by commercial purposes, some of which are the production of radioactive sources in cancer treatment and medical imaging, sterilizing in medical hardware and food, energetic radioisotope production, production of semiconductors for electronic systems, etc. [3–8].

For those purposes, different accelerators as linear and circular types are available all over the world. A linear accelerator (linac) includes merely accelerating structures as the charged particles do not need to be rotated, and linacs exclusively utilize from one acceleration pass. This situation leads to increasing the length of the accelerator to reach high-energy levels. To accelerate the ions, linacs have an accelerating tube including a number of electrodes, where a high-frequency alternating voltage and ions are accelerated in the gap between the electrodes in suitable voltage [1, 2], for example, Stanford Linear Accelerator (SLAC), which is a 3.2-km-long electron-positron collider [9], and the proton accelerator with 800 MeV in the Los Alamos Neutron Science Center (LANSCE) [10] as shown in **Figures 1** and **2**.

On the other hand, the particles in circular accelerators rotate the same circuit to reach energy range that are wanted due to getting an energy boost at each turn. However, to keep particles in their circular orbit, powerful magnetic field must be provided by device [1] such as cyclotron and synchrotron. Cyclotrons are the well-known and the most successful accelerators. They accelerate the ions by a radio-frequency (RF) electrical field and have a magnetic field to constrain the ions to move in spiral path. The cyclotrons are used in nuclear medicine and radioisotope productions (**Figure 3**) [2].

The ion energies accelerated in cyclotron do not reach high-energy levels and do not have suitable structure to collide two particles. However, a synchrotron can explore higher and higher energies and can become larger and larger because the radius of an accelerator is connected to the energy, feasibility, and cost. The synchrotrons may overcome the energy limitations of cyclotrons. The acceleration

Figure 1.
Stanford linear accelerator (SLAC) at Stanford University [9].

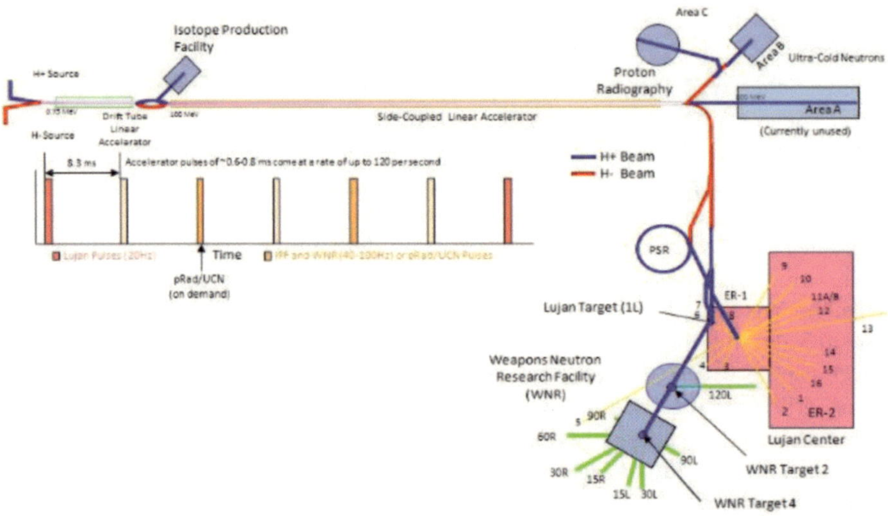

Figure 2.
The structure of a linac at the Los Alamos Neutron Science Center (LANSCE) [10].

in the synchrotrons is performed by a RF oscillator [2]. The most powerful synchrotron in the world is Large Hadron Collider (LHC), which is a collider accelerator creating center-of-mass energies of 13 TeV, in the European Organization for Nuclear Research (CERN). Colliders that can accelerate two ions and cause them to collide head on are particle accelerators. LHC and many particle accelerators in the CERN are presented in **Figure 4** as well as using particle types [12].

In addition to LHC, the Future Circular Collider (FCC) in the CERN which was started in 2013 includes the design of a high-luminosity e^+e^- collider (FCC-ee) which is a possible first step, as well as a proton-electron collision option (FCC-he). The time lines of developing FCC and LHC circular colliders at CERN are demonstrated in **Figure 5** [13].

Additionally, new accelerator types in the design and construction of a new multipurpose research plant have been built by various countries, one of which is Belgian Nuclear Research Center (SCK.CEN) in Mol, Belgium, where the *multi-purpose hybrid research reactor for high-tech applications (MYRRHA) is* a versatile research infrastructure and the world's first research reactor driven by a particle accelerator [14].

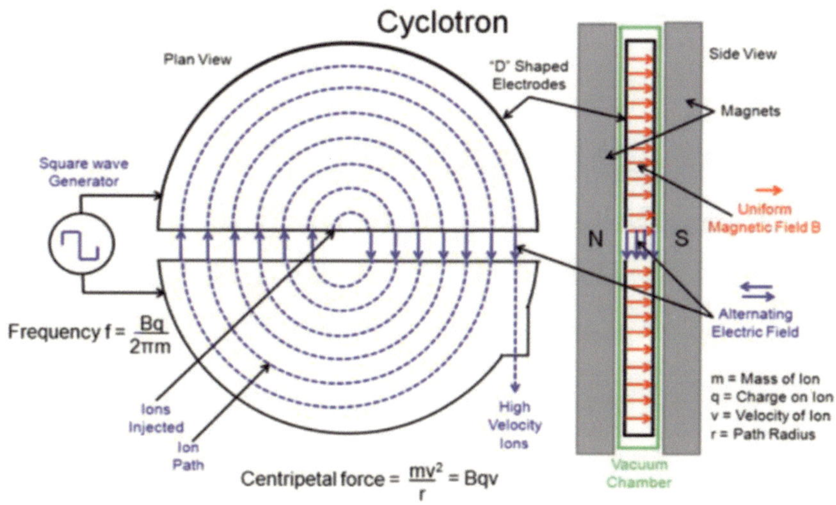

Figure 3.
The sections of a cyclotron structure [11].

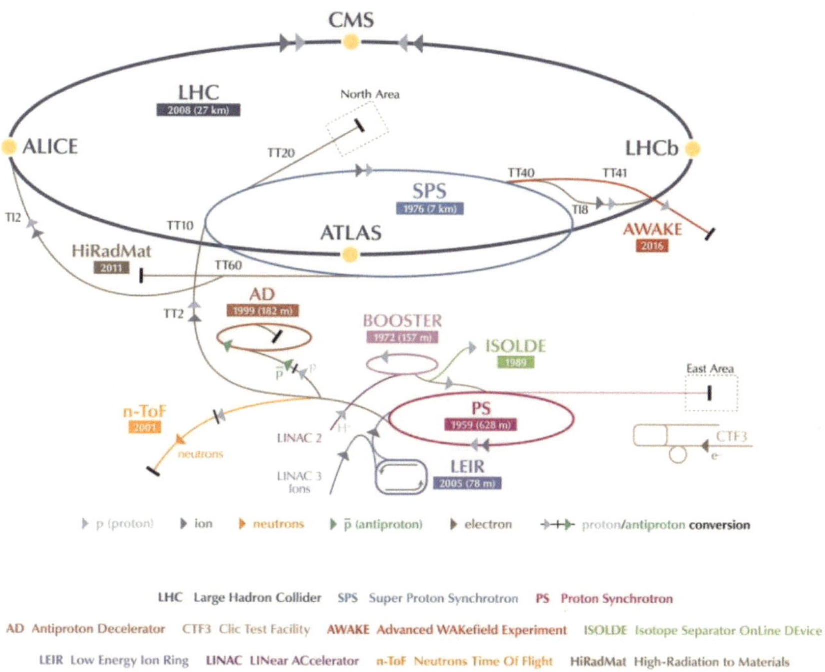

Figure 4.
An overview of LHC and the other accelerators at CERN and particle types used in accelerators [12].

MYRRHA project that is a candidate for European Investment Bank financing is supported by the Belgian Federal Government, based on a total budget of € 1.6 bn, and the government committed € 558 m to the project's phased approach. MYRRHA has a 400-m-long linear accelerator to provide protons at an energy of 600 MeV to the reactor system which is unique in many aspects because it is a pool-type reactor cooled by 7800 tons of lead-bismuth eutectic (LBE) and the reactor system involves the double-walled 16-m-high and 10-m-wide vessel [15, 16] as shown in **Figure 6**.

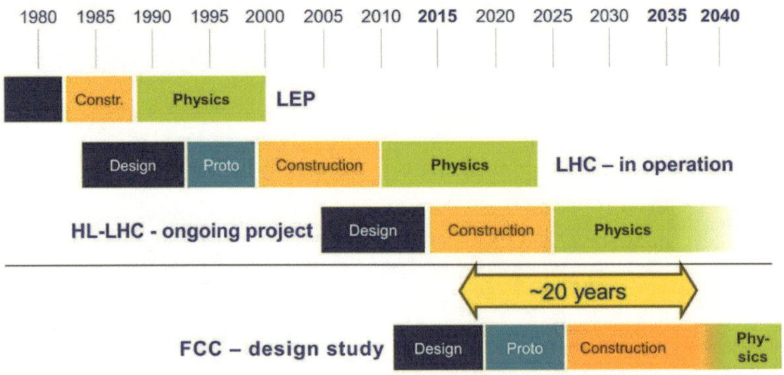

Figure 5.
The past, present, and future colliders and their design studies based on time [13].

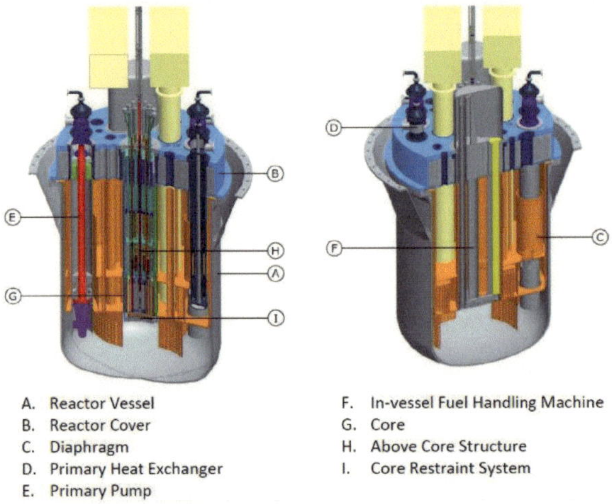

Figure 6.
MYRRHA reactor and sections at SCK.CEN [15].

MYRRHA is constructed in three phases (**Figure** 7). The first phase is the design and construction of the first linac section (up to 100 MeV) which is scheduled for completion in 2026. The second phase is the extension of the 100 MeV linac to 600 MeV up to 2033, where the linac will be approximately 400 m long. The third phase is the construction of the reactor, and the reactor is scheduled to be commissioned in 2036. Furthermore, the project involves the proton target facility (PTF) and the fusion target station (FTS) sections [17, 18].

This book mentions about important applications of accelerators and colliders and some key novelties in accelerator technology. Section two refers to the production of medical Ir-192 used in brachytherapy on Os-targets in the irradiation time of 1 h with constant beam current of 1 μA for the energy range of 1–100 MeV via particle accelerators. The third section explains a concise overview of key issues of vacuum system design and operations in high-energy, high-intensity, and high-luminosity accelerators and colliders. The fourth section involves a novel scheme of a gas-filled capillary accelerator driven by a laser pulse formed from two-mode mixing of the capillary eigenmodes and proposes the laser-plasma linear collider. The next sections include theoretical calculations of the masses of the elementary fermions and a review of the elementary particles, quantum chromodynamics (QCD), and strong interactions in QCD theory.

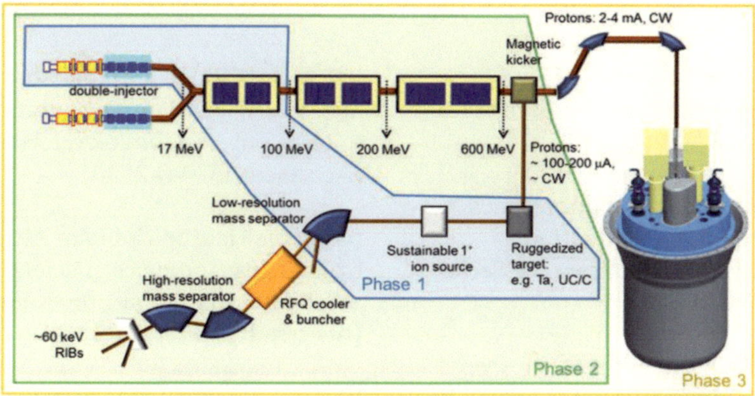

Figure 7.
MYRRHA's three phases at SCK.CEN [17].

Author details

Ozan Artun
Department of Physics, Zonguldak Bülent Ecevit University, Zonguldak, Turkey

*Address all correspondence to: ozanartun@beun.edu.tr; ozanartun@yahoo.com

References

[1] Accelerators. Available from: https://home.cern/science/accelerators [Accessed: 25 May 2020]

[2] Pagliarone CE. Collider physics an experimental introduction. Journal of Physics: Conference Series. 2011;**287**:012005

[3] Artun O. Production of polonium-208, 209 and 210 for use in nuclear battery via particle accelerator. Applied Physics A. 2020;**126**:386

[4] Artun O. Investigation of the production of promethium-147 via particle accelerator. Indian Journal of Physics. 2017;**91**(8):909-914

[5] Artun O. Investigation of the production of cobalt-60 via particle accelerator. Nuclear Technology & Radiation Protection. 2017;**32**(4):327-333

[6] Artun O. Investigation of production of samarium-151and europium-152,154,155 via particle accelerator. Modern Physics Letters A. 2019;**34**(20):1950154

[7] Artun O. Investigation of production of medical 82Sr and 68Ge for 82Sr/82Rband 68Ge/68Ga generators via proton accelerator. Nuclear Science and Techniques. 2018;**29**:137

[8] Artun O. Estimation of the production of medical Ac-225 on thorium material via proton accelerator. Applied Radiation and Isotopes. 2017;**127**:166-172

[9] SLAC. Available from: https://www6.slac.stanford.edu/ [Accessed: 25 May 2020]

[10] LANSCE. Available from: https://lansce.lanl.gov/ [Accessed: 25 May 2020]

[11] About Cyclotron. Available from: https://www.assignmentpoint.com/science/physics/about-cyclotron.html [Accessed: 25 May 2020]

[12] Large Hadron Collider. Available from: https://home.cern/science/accelerators/large-hadron-collider [Accessed: 25 May 2020]

[13] Zimmermann F. Future colliders for particle physics "big and small". Nuclear Instruments and Methods in Physics Research A. 2018;**909**:33-37

[14] MYRRHA. Available from: https://www.sckcen.be/en/projects/myrrha [Accessed: 25 May 2020]

[15] De Bruyn D, Abderrahim HA, Baeten P, Leysen P. The MYRRHA ADS project in Belgium enters the front end engineering phase. Physics Procedia. 2015;**66**:75-84

[16] MYRRHA Accelerator. Available from: https://www.myrrha.be/myrrha-project/myrrha-accelerator/ [Accessed: 25 May 2020]

[17] ISOL@MYRRHA: Fundamental Physics at MYRRHA. Available from: https://isolmyrrha.sckcen.be/ [Accessed: 25 May 2020]

[18] MYRRHA Phased Implementation. Available from: https://www.myrrha.be/myrrha-project/myrrha-phased-implementation/ [Accessed: 25 May 2020]

Chapter 2

Investigation of the Production of Medical Ir-192 Used in Cancer Therapy via Particle Accelerator

Ozan Artun

Abstract

To investigate the production of medical Ir-192 radionuclide used in brachytherapy on Os targets in the energy range of $E_{particle} = 100 \rightarrow 1$ MeV, we calculated the cross-section results for charged particle-induced reactions. The calculation was done via TALYS code and simulated activity and yield of product of each reaction process in the irradiation time of 1 h with constant beam current of 1 μA. The calculated results were compared with experimental data in the literature. Moreover, based on the calculated cross-section data and the mass stopping powers obtained from X-PMSP program, the integral yield results of all the reaction processes to produce Ir-192 on Os targets were presented as a function of incident particle energy. The obtained results were discussed to recommend appropriate reaction processes and targets for the production of Ir-192.

Keywords: particle accelerator, cancer therapy, radioisotope production, yield, irradiation

1. Introduction

Particle accelerators and colliders are widespread ranging from radioisotope production for medical subjects to exploration of new particles in the high physics. For this aim, different collider and accelerator systems have been established by research intuitions and organizations across the world based on powerful accelerators. In general, the largest particle accelerators (in TeV) are used for elementary particle physics to research subatomic particles in physics such as Fermilab or Large Hadron Collider (LHC) at CERN. Additionally, investigating the structure of nuclei in GeV and researching isotopes in nuclear physics have been performed by Relativistic Heavy Ion Collider (RHIC) at the Brookhaven National Laboratory and Los Alamos Neutron Science Center (LANSCE) at Los Alamos. At lower energies, accelerators used in medicine have vital importance in terms of treatment of cancer and diagnosis aims of cancer tissues, especially for the production of medical radioisotopes in MeV. That's why, for an application of particle accelerators, this chapter gives the production of medical iridium-192 used in brachytherapy via particle accelerator.

The iridium-192 that can be used in brachytherapy with interstitial implantation by irradiating malignant tumors is important for therapeutic aims especially for the brain, uterus, head, etc. [1] because the radioisotope Ir-192 with low-energy and high-intensity beta radiation has effective decay properties, which are $T_{1/2} = 78.83$ d,

IntechOpen

EC = 5%, I_{β}^{-} = 95%, and E_{β}^{-} = 7 MeV, for using in medical applications [2]. Therefore, the production of such a radioisotope is fairly attractive, and there are some production studies of Ir-192 in literature by Langille et al. [3], Szelecsényi et al. [4], Tarkanyi et al. [5], and Hilger et al. [2]. These studies mainly intensify the production of Ir-192 on enriched Os-192 targets via (p, n) reaction processes, and the measurement by Tarkanyi et al. [5] only was a different reaction process, including ^{192}Os(d,2n)^{192}Ir together with isomeric states, $^{192m1+g}$Ir. The production of Ir-192 is also available for different methods such as the (neutron, gamma) reaction process in nuclear reactors [2, 5–7]. However, there are six stable isotopes of Os in nature, namely, Os-184 (0.02%), Os-187 (1.96%), Os-188 (13.24%), Os-189 (16.15%), Os-190 (26.26%), and Os-192 (40.78%) [8].

Therefore, in addition to ^{192}Os(p, n)^{192}Ir reaction, we take into account induced reactions with deuteron, triton, helium-3, and alpha particles in the energy range between 1 MeV and 100 MeV [9–13]. Furthermore, we calculated and simulated the production of Ir-192 on Os-192, Os-189, and Os-190 targets for each charged particle. The cross-section and the integral yield calculations were performed by TALYS 1.9 code [14] and X-PMSP 2.0 program [12, 15–18], which is used to obtain the mass stopping power results of Os for the charged particles. The activities and the yield of products of reaction processes were carried out under certain circumstances, such as current beam of 1 μA and irradiation time of 1 h. The calculated results were compared with the experimental data obtained from Exfor database [19] and other works in the literature. It is obvious that the obtained data are important to determine the production of Ir-192 in terms of other targets and the charged particle-induced reactions due to providing new nuclear data for its production.

2. Materials and methods

Here, we have investigated the production of radioisotope Ir-192 used in brachytherapy through nuclear reaction processes on enriched stable Os target isotopes. For this aim, we calculated the cross-section and the integral yield curves in $E_{particle} = 100 \rightarrow 1$ MeV energy region for each reaction process. Moreover, the activities and yields of product of reaction processes as a function of irradiation time are simulated under certain conditions via TALYS 1.9 code. Besides proton-induced reaction, the radioisotope Ir-192 can be produced by deuteron, triton, helium-3, and alpha particles accelerated by particle accelerators on natural Os targets in 1–100 MeV energy range.

To produce Ir-192 on osmium targets (^{189,190,192}Os), we followed out all reaction processes under certain situations to estimate integral yield, activity, and yield of product curves for the charged particle-induced reactions based on incident particle energy region between 1 MeV and 100 MeV. We assumed the purity of osmium targets as being above 99%, the target areas are about 1 cm^2, and the target thicknesses are of uniform density during the irradiation processes in reactions. Moreover, the effective thicknesses of targets can change from 0.065 (^{189}Os for alpha-induced reaction) cm to 0.704 (^{192}Os for proton-induced reaction) as dependent on reaction process. For each reaction process, the produced heat in targets is about 0.099 kW, and the density of target materials is 22.600 g/cm^3.

Furthermore, the target materials are bombarded by charged particles accelerated in particle accelerator with beam current of 1 μA during irradiation time of 1 h in energy region $E_{particle} = 100 \rightarrow 1$ MeV, and the cooling time is in 24 h. All the reaction processes never involve loss in activity and yield during irradiation time. The activities, yields of product, and integral yields are performed by conditions mentioned above.

The process shows that when the Os targets are bombarded by the charged particles accelerated in particle accelerators, the residual nuclei are made up, and it turns into Pt^* and Ir^*(* means compound nucleus) isotopes by emitting different particles in very small time interval, and afterwards, the radioisotope Ir-192 is formed. Therefore, the nuclear structure properties of Pt and Ir isotopes in nuclear reaction process are also calculated such as proton, deuteron, and triton separation energies. To determine neutron (S_n), proton (S_p), deuteron (S_d), and triton (S_t) separation energies of Pt and Ir isotopes, we exploited TALYS code 1.9, which included different theoretical [20, 21] experimental tables [22] and Duflo-Zuker mass formula [14] to calculate the mass values of nuclei.

Based on reaction processes used in this work, PEQ reaction mechanism can be suitably explained by two-component exciton model which can make distinction of particle (p) and hole (h) pairs in reaction flow, besides neutron (ν) and proton (π) particles [14]:

$$\frac{d\sigma_k^{PE}}{dE_k} = \sigma^{CF} \sum_{p_\pi=p_\pi^0}^{p_\pi^{max}} \sum_{p_\nu=p_\nu^0}^{p_\nu^{max}} W_k(p_\pi, h_\pi, p_\nu, h_\nu, E_k)\tau(p_\pi, h_\pi, p_\nu, h_\nu) \times P(p_\pi, h_\pi, p_\nu, h_\nu) \quad (1)$$

where σ^{CF}, P, τ, and W_k are compound formation cross-section, pre-equilibrium population, mean lifetime, and emission rate, respectively. It is important to understand the estimation of particle emission situations which is available if the exciton number equals to three (n = 2p1h) minimally. On the other hand, for an ejectile k, the emission rate W_k as dependent on two-component particle hole state density, ω is given by [14, 23, 24]

$$W_k(p_\pi, h_\pi, p_\nu, h_\nu, E_k) = \frac{2s_k+1}{\pi^2\,^3}\mu_k E_k \sigma_{k,inv}(E_k) \times \frac{\omega(p_\pi - Z_k, h_\pi, p_\nu - N_k, h_\nu, E^{tot} - E_k)}{\omega(p_\pi, h_\pi, p_\nu, h_\nu, E^{tot})} \quad (2)$$

where $\sigma_{k,inv}(E_k)$, Z_k, μ_k, and s_k are the inverse cross-section, the charge number of the ejectile, the relative mass, and spin, respectively. Additionally, the ω in Eq. (2) is given by single-particle state densities (g_π,g_ν) [24, 25]:

$$\omega(p_\pi, h_\pi, p_\nu, h_\nu, E_x) = \frac{g_\pi^{n_\pi} g_\nu^{n_\nu}}{p_\pi! h_\pi! p_\nu! h_\nu! (n-1)!}\left(U - A(p_\pi, h_\pi, p_\nu, h_\nu)\right)^{n-1} \times f(p, h, U, V) \quad (3)$$

where f, $P_{p,h}$, and A represent the finite well function, Fu's pairing function, and the Pauli correction as seen in Eqs. 4–6 and $U = E_x - P_{p,h}$ [26]:

$$P_{p,h} = \Delta - \Delta\left[0.996 - 1.76\left(\frac{n}{n_{crit}}\right)^{1.6} / \left(\frac{E_x}{\Delta}\right)^{0.68}\right]^2 \quad (4)$$

Eq. (4) is valid if $E_x/\Delta \geq 0.716 + 2.44\left(\frac{n}{n_{crit}}\right)^{2.17}$; otherwise Eq. (4) equals to Δ.

$$A(p_\pi, h_\pi, p_\nu, h_\nu) = \frac{[max\,(p_\pi, h_\pi)]^2}{g_\pi} + \frac{[max\,(p_\nu, h_\nu)]^2}{g_\nu} - \frac{p_\pi^2 + h_\pi^2 + p_\pi + h_\pi}{4g_\pi} - \frac{p_\pi^2 + h_\pi^2 + p_\pi + h_\pi}{4g_\nu} \quad (5)$$

$$f(p,h,E_x,V) = 1 + \sum_{i=1}^{h} (-1)^i \binom{h}{i} \left[\frac{E_x - iV}{E_x}\right]^{n-1} \Theta(E_x - iV) \qquad (6)$$

where Θ and V in the finite well function are unit step function and the potential well depth [27, 28].

3. Results and discussions

Based on charged particle-induced reaction processes, the production of Ir-192 on ^{189,190,192}Os target materials via particle accelerator in energy region between 1 MeV and 100 MeV, we investigated the cross-section calculations (**Figure 1**), the activity (**Figure 2**), and the yield of product (**Figure 3**) curves for all the reaction processes, as well as the separation energies (**Figure 4**). Additionally, the integral yield results are calculated by data of the cross-section calculations in **Figure 1** and mass stopping power results in **Figure 5** obtained from X-PMSP 2.0 program. The separation energies in **Figure 4** are calculated to understand particle emissions that divide from residual nuclei (^{191,192,193,194,195}Ir* and ^{192,193,194,195,196}Pt*), which is composed of bombardment of Os target.

3.1 Cross-section calculations for the production of Ir-192

To produce Ir-192, the cross-section curves with proton-, deuteron-, triton-, he-3-, and alpha-induced reactions on ^{189,190,192}Os targets are presented by **Figure 1** in 1–100 MeV energy range together with experimental data by Szelecsényi et al. [4], Tarkanyi et al. [5], and Hilgers et al. [2]. The measurements by Szelecsényi et al. [4] and Hilgers et al. [2] were only performed by proton-induced reaction on Os-192 target up to 66 MeV. On the other hand, the data reported by Tarkanyi et al. [5] reaches 20 MeV deuteron incident energy only on Os-192 target. It is obvious

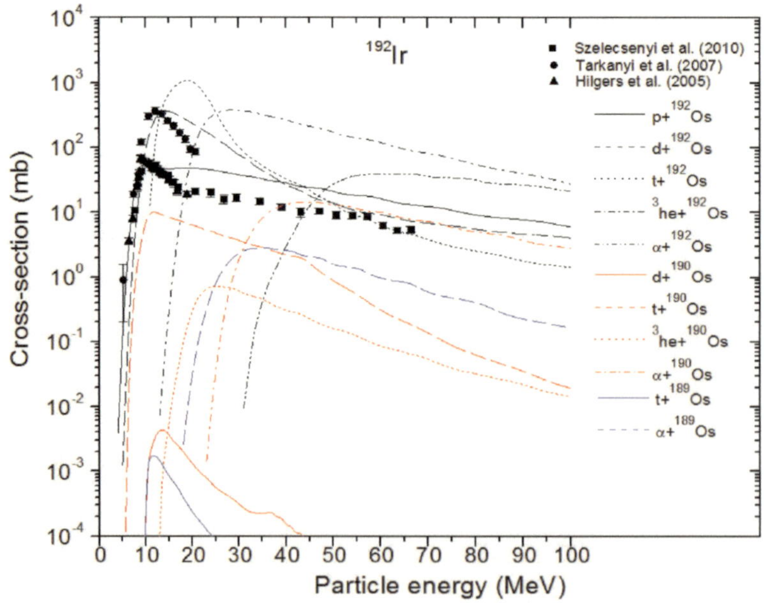

Figure 1.
Cross-section curves for the production of Ir-192.

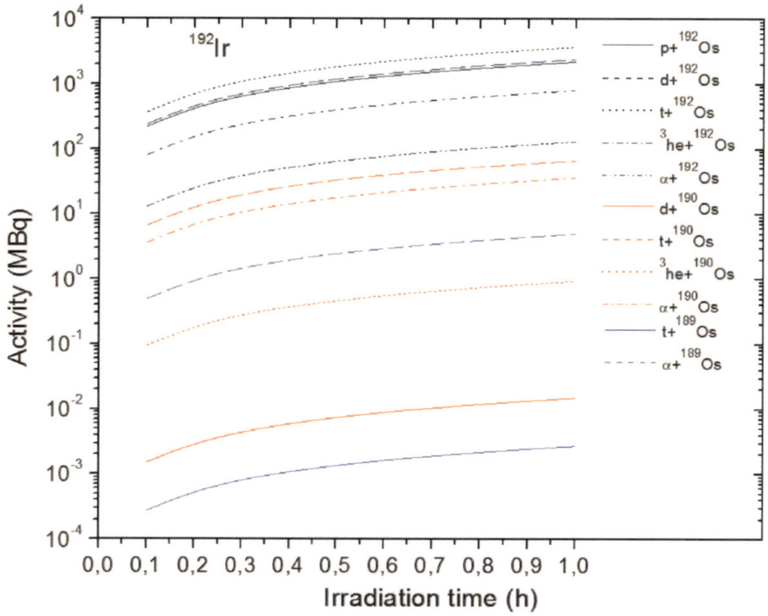

Figure 2.
Activity curves for the production of Ir-192 as a function of irradiation time.

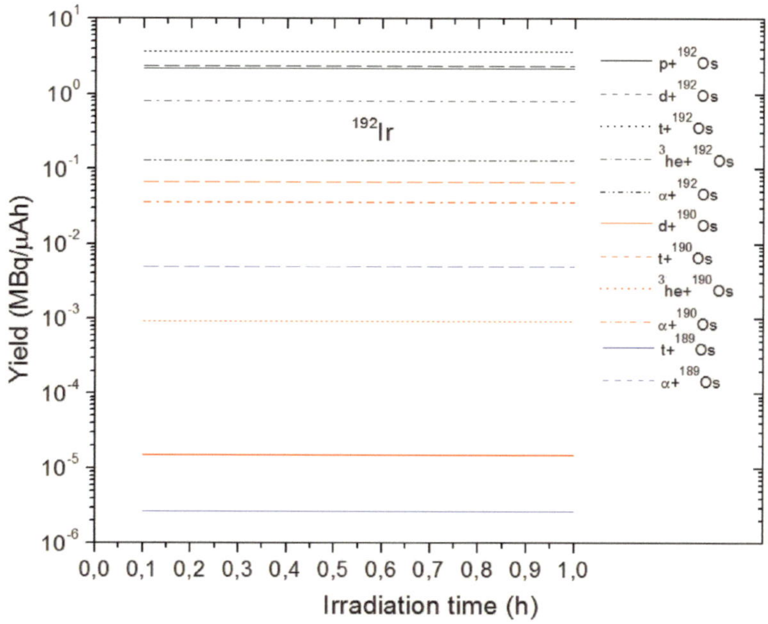

Figure 3.
Yield of product curves for the production of Ir-192 as a function of irradiation time.

that the production of Ir-192 for the triton-, he-3-, and alpha-induced reactions Os-192 target does not involve any experimental result or theoretical result. In addition to Os-192 target, for Os-190 and Os-189 targets, unfortunately, the cross-section results for the production of Ir-192 are not come across different works. In comparisons of the calculated cross-section and the experimental data for Os-192 target, the theoretical result for proton-induced reaction is consistent with that of Szelecsényi et al. [4] and Hilgers et al. [2] from threshold energy to the maximum

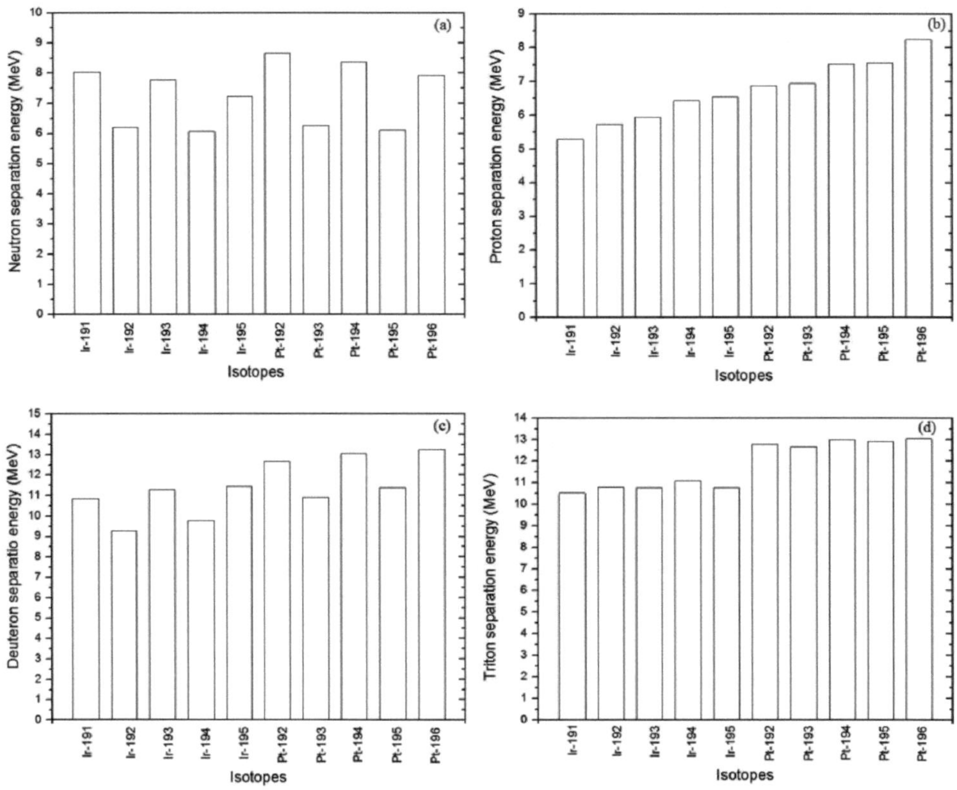

Figure 4.
Calculation of separation energies for some Ir and Pt isotopes in reaction process.

cross-section value. Thus, we can conclude that the theoretical cross-section curves are consistent with the experimental data up to the highest value of cross-section curves. Moreover, the data reported by Tarkanyi et al. [5] are generally in good agreement with theoretical result up to 11 MeV deuteron incident energy. Besides proton- and deuteron-induced reaction on Os-192 target, it is important to note here that the cross-section values of the other induced reactions are suitable to obtain the radioisotope because while the maximum cross-section values for proton- and deuteron-induced reactions reach ~54 mb and ~371 mb at 9 MeV and 14 MeV, the cross-section values of the triton-, he-3-, and alpha-induced reactions ramp up to ~1100 mb, ~383 mb, and ~40 mb at 19 MeV, 29 MeV, and 59 MeV, respectively. Furthermore, for Os-189 and Os-190 targets, alpha-induced reactions are clearly suitable compared to the other reaction types.

The results show that the production of the radioisotope Ir-192 can be produced by triton-, he-3-, and alpha-induced reactions, in addition to proton- and deuteron-induced reactions, and such productions are valid in less energy than that of 100 MeV. Therefore, to well understand the production of Ir-192 for each reaction process, we estimated activity and yield curves as a function of irradiation time by simulating under certain conditions.

3.2 Simulation of activity and yield of product for the production of Ir-192

The calculated cross-section results are in good agreement with the available experimental data in the literature. The reaction processes are simulated to determine the activity and yield of product curves for the production of Ir-192 as shown

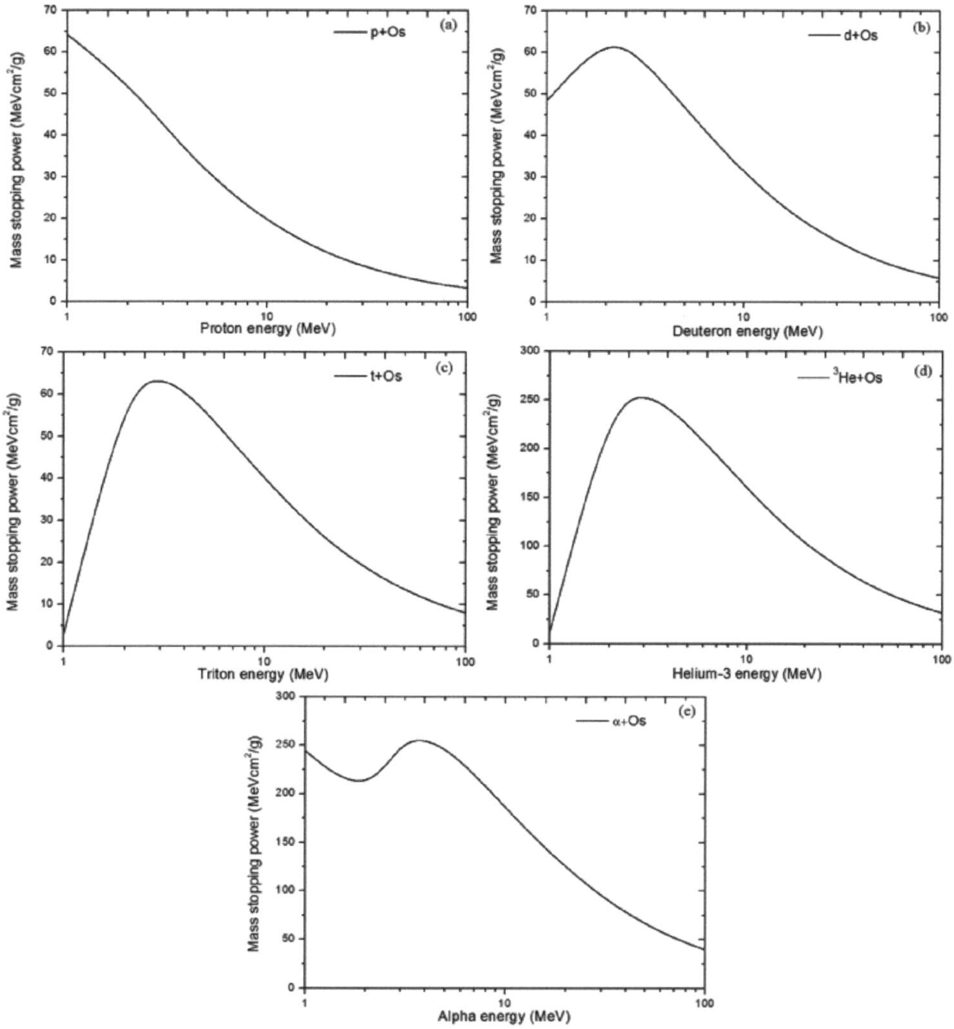

Figure 5.
Mass stopping power for the charged particles on Os target.

in **Figures 2** and **3**. When compared with the cross-section curves, the activity and yield of product results are consistent with the cross-section curves. For Os-192 target, at end of irradiation time, the triton-induced reaction has the highest activity value of 3690 MBq. The deuteron- and proton-induced reaction reach almost the same activity values, such as 2390 MBq and 2190 MBq, respectively. On the other hand, helium-3- and alpha-induced reaction processes also have high activity values, 803 MBq and 130 MBq.

It is clear that the proton- and deuteron-induced reactions have higher activity and yield values than those of alpha- and he-3-induced reactions; however, this can be an advantage rather than disadvantage. Ir-192 is a radioisotope for brachytherapy especially for vitals, and it is used inside the capsule which is settled next to the tumor region to burn out cancerous tissues, and the radioisotope with Ir-192 high activity can bring about harm for healthy tissues. Therefore, the reaction processes with the alpha and he-3 particles can be more suitable in terms of application to patients compared to the production of Ir-192 with proton- and deuteron-induced reactions.

In addition to the production of Ir-192 on Os-192 target via the charged particle-induced reactions, when considered Os-190 and Os-189 targets, the alpha-induced reactions for both targets give suitable activity values, 35.9 MBq and 4.92 MBq, respectively. Moreover, the activity value of triton-induced reaction on Os-190 target also reaches 66.2 MBq; however, it is obvious that the triton-induced reaction on Os-189 target is insufficient. Unfortunately, it is important to say that we do not recommend the radioisotope production with triton-induced reactions due to expensive obtaining of triton. Therefore, for the production of Ir-192 on Os-190 and Os-189 targets, we can dependably propose the alpha-induced reaction process.

3.3 Calculation of separation energies

When the ^{189,190,192}Os targets are bombarded by proton, deuteron, triton, helium-3, and alpha particles accelerated by particle accelerator, new excited compound nuclei, ^{191,192,193,194,195}Ir and ^{192,193,194,195,196}Pt may be made up, and after then too very while, these excited nuclei turn into new residual nuclei by emitting particles such as proton, neutron, deuteron, etc.

One of the residual nuclei also is the radioisotope Ir-192; however, there are different ways to produce Ir-192, for example, the compound nucleus Ir-193 changes to Ir-192 by emitting a neutron, and if the compound nucleus Pt-193 spreads out a proton or the Pt-194 emits a deuteron, then the radioisotope Ir-192 may be produced. Therefore, to form Ir-192 on ^{189,190,192}Os targets, we calculated the separation energies of possible compound nuclei as shown in **Figure 4**. When considering the calculated separation energies for isotopes of Pt, they are higher than those of the isotopes of Ir. The he-3 and alpha separation energies are not added due to the formation of Pt and Ir isotopes in reaction processes via the charged particles. Moreover, it is obvious that the neutron separation and proton separation energies are lower than those of the other separation energies.

3.4 Calculation of integral yield of reaction processes

To peruse the production of Ir-192 on ^{189,190,192}Os targets for the charged particle-induced reactions in energy region between 1 MeV and 100 MeV, we calculated the integral yield results of each reaction process via the calculated cross-section results in **Figure 1** and the stopping powers of Os for the charged particles obtained from X-PMSP 2.0 program (**Figure 5**); the obtained integral yield curves are presented in **Figure 6** as a function of particle energy. Based on the simulation conditions of activity and yield of product of reaction processes, the integral yield calculations were carried through in irradiation time of 1 h and a constant beam current of 1 μA. The calculated integral yield curves were taken by the optimization of a production route into account by providing high radionuclidic purity of the Ir-192, including minimum impurity level and maximum yield value of the product.

When taking into account the nuclear reaction processes on Os-192 target, it is clearly stated that the proton-induced reaction has only experimental yield data reported by Szelecsényi et al. [4] up to 30 MeV incident proton energy. The calculated integral yield curve for proton-induced reaction is generally consistent with measurement by Szelecsényi et al. [4], up to ~15 MeV; however, beyond 15 MeV, there is a little bit of difference when compared to the theoretical result. The reason of such a distinction is why there is difference of both theoretical and experimental cross-section curves in **Figure 1** and the error rates in the theoretical calculations and the experimental processes. Even the theoretical integral yield curve is close to the experimental data, and in terms of analyzing accuracy of the integral yield

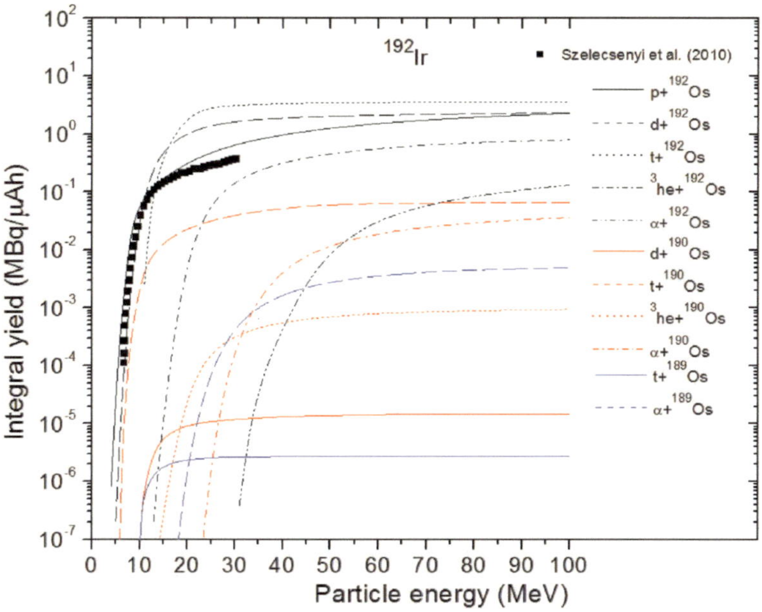

Figure 6.
Integral yield curves for the production of Ir-192 as a function of incident particle energy.

curves, it can clearly be said that the results are believable and dependable. On the other hand, the other induced reactions to produce Ir-192 are suitable, especially for deuteron- and triton-induced reactions which are of higher integral yield values (2 MBq/μAh and 4 MBq/μAh) than the proton-induced reaction (1 MBq/μAh) behind 30 MeV. Moreover, the He-3 induced reaction process gives good results; however, its appropriate energy range reaches to higher energy compared to the other induced reactions, and the alpha-induced reaction has the highest energy range to produce Ir-192.

Furthermore, based on the theoretical results, it is important to note here that while the energy range for the production of Ir-192 through proton-induced reaction can be determined as 20 → 8 MeV, the appropriate energy ranges of triton- and deuteron-induced reactions are 22 → 12 MeV and 21 → 9 MeV, respectively. These energy ranges can be formed higher than 70% of the radioisotope Ir-192; however, in these energy ranges, the contaminations in the production of Ir-192 would be brought about. The production of Ir-192 in proton-induced reactions can include the contamination of the radioisotope Ir-190 higher than 16 MeV incident proton energy. From the point of view of Szelecsényi et al. [4], such as a circumstance is valid, they pointed out that the contamination of Ir-190 for proton-induced reaction reaches 70% at 17 MeV, and unfortunately, the $^{190}Ir/^{192}Ir$ activity ratio keeps up fairly high [4].On the other hand, the production of Ir-192 on Os-190 and Os-189 targets are also presented in **Figure 6** as a function of particle energy. For Os-190 target, the triton reaction and alpha-induced reactions are more suitable than those of the other reactions. The alpha reaction process gives the best integral yield result for Os-189 target.

4. Conclusion

This paper mainly investigated the production of Ir-192 used in brachytherapy via $^{189,190,192}Os$ targets bombarded by proton, deuteron, triton, helium-3, and alpha particles accelerated in particle accelerators, instead of its production in nuclear reactors. Therefore, we calculated the cross-section curves and simulated activities and yields of product of all reaction processes. Additionally, to determine the energy ranges for the formation of Ir-192 in nuclear reaction processes, we estimated the integral yield results for each reaction process in energy region between 1 MeV and 100 MeV.

The calculated data were compared with the experimental data in the literature, and the obtained results for proton-induced reaction on Os-192 target showed that the theoretical results were in good agreement with available experimental data based on the cross-section and the integral yield results. When taking conformity of proton-induced reaction results into account, it is important to note here that the obtained data in this work put dependable results for the production of Ir-192 forward the literature. Because, in addition to proton-induced reaction, the obtained data for the production of Ir-192 contributes to the literature in terms of deuteron-, triton-, helium-3-, and alpha-induced reaction processes. Moreover, the calculations and simulations were also performed by charged induced reaction processes on Os-190 and Os-189 targets besides Os-192 target.

Based on the obtained results, in addition to the production of proton-induced reaction on Os-192 target, it can be said that deuteron-, triton-, helium-3-, and alpha-induced reaction on Os-192 target are suitable to produce Ir-192 in energy region between 1 MeV and 100 MeV. These energy ranges for the production of Ir-192 irrespective of contamination of the other product in the reaction processes can be given for proton-, deuteron-, triton-, he-3-, and alpha-induced reactions: $20 \rightarrow 8$ MeV, $21 \rightarrow 9$ MeV, $22 \rightarrow 12$ MeV, $35 \rightarrow 19$ MeV, and $65 \rightarrow 34$ MeV, respectively. Furthermore, in the case of Os-190 target, the triton reaction and alpha-induced reactions are suitable compared to the other reactions, and the energy ranges are given as $20 \rightarrow 7$ MeV and $55 \rightarrow 25$ MeV. For Os-189 target, the alpha-induced reaction process can be proposed for the production of Ir-192 in energy range $50 \rightarrow 21$ MeV.

Author details

Ozan Artun
Department of Physics, Zonguldak Bülent Ecevit University, Zonguldak, Turkey

*Address all correspondence to: ozanartun@beun.edu.tr; ozanartun@yahoo.com

References

[1] Ghiassi-Nejad M, Jafarizadeh M, Ahmadian-Pour MR, Ghahramani AR. Dosimetric characteristics of ^{192}Ir sources used in interstitial brachytherapy. Applied Radiation and Isotopes. 2001;**55**:189-195

[2] Hilgers K, Sudar S, Qaim SM. Experimental study and nuclear model calculations on the ^{192}Os[p,n]^{192}Ir reaction: Comparison of reactor and cyclotron production of the therapeutic radionuclide ^{192}Ir. Applied Radiation and Isotopes. 2005;**63**:93-98

[3] Langille G, Yang H, Zeisler SK, Hoehr C, Storr T, Andreoiu C, et al. Low energy cyclotron production and cyclometalation chemistry of iridium-192. Applied Radiation and Isotopes. 2016;**115**:81-86

[4] Szelecsényi F, Vermeulen C, Steyn GF, Kovács Z, Aardaneh K, van der Walt TN. Excitation functions of ^{186,187,188,189,190,192}Ir formed in proton-induced reactions on highly enriched 192Os up to 66 MeV. Nuclear Instruments and Methods B. 2010;**268**: 3306-3314

[5] Tarkanyi F, Hermanne A, Takacs S, Hilgers K, Kovalev SF, Ignatyuk AV, et al. Study of the ^{192}Os[d,2n] reaction for production of the therapeutic radionuclide ^{192}Ir in no-carrier added form. Applied Radiation and Isotopes. 2007;**65**:1215-1220

[6] Ananthakrishnan M. Iridium 192. In: Manual for Reactor Produced Radioisotopes IAEA-TECDOC-1340. 2003. p. 116. Available from: https://www.pub.iaea.org/MTCD/Publications/PDF/te_ 1340_web.pdf [Accessed: 02 January 2020]

[7] Schaeken B, Vanneste F, Bouiller A, Hoornaert MT, Vandenbrooeck S, Hermans J, et al. ^{192}Ir brachytherapy sources in Belgian hospitals. Nuclear Instruments and Methods A. 1992;**312**: 251-256

[8] NIST, National Institute of Standards and Technology. 2018. Available from: https://www.nist.gov/ [Accessed: 02 January 2020]

[9] Artun O. Calculation of productions of PET radioisotopes via phenomenological level density models. Radiation Physics and Chemistry. 2018; **149**:73-83

[10] Artun O. Investigation of the productions of medical ^{82}Sr and ^{68}Ge for ^{82}Sr/^{82}Rb and ^{68}Ge/^{68}Ga generators via proton accelerator. Nuclear Science and Techniques. 2018;**29**:137-147

[11] Artun O. Investigation of the production of cobalt-60 via particle accelerator. Nuclear Technology & Radiation Protection. 2017;**32**(4): 327-333

[12] Artun O. Estimation of the production of medical Ac-225 on thorium material via proton accelerator. Applied Radiation and Isotopes. 2017; **127**:166-172

[13] Artun O. Investigation of the production of promethium-147 via particle accelerator. Indian Journal of Physics. 2017;**91**:909-914

[14] Koning A, Hilaire S, Goriely S. Talys Manual 1.9. 2017. Available from: http://www.talys.eu/download-talys/ [Accessed: 02 January 2020]

[15] Artun O. Calculation of the mass stopping powers of medical, chemical, and industrial compounds and mixtures. Nuclear Technology & Radiation Protection. 2018;**33**(4):356-362

[16] Artun O. X-particle the mass stopping. In: Power, X-PMSP

Version 2.00. 2018. Available from: https://www.x-pmsp.com/⟩ [Accessed: 02 January 2020]

[17] Artun O. Calculation of productions of medical ^{201}Pb, ^{198}Au, ^{186}Re, ^{111}Ag, ^{103}Pd, ^{90}Y, ^{89}Sr, ^{77}Kr, ^{77}As, ^{67}Cu, ^{64}Cu, ^{47}Sc and ^{32}P nuclei used in cancer therapy. Applied Radiation and Isotopes. 2019;**144**:64-79

[18] Artun O. Investigation of production of samarium-151 and europium-152,154,155 via particle accelerator. Modern Physics Letters A. 2019;**33**:1950154

[19] Exfor. Available from: https://www.nds.iaea.org/exfor/exfor.htm; 2019

[20] Goriely S, Pearson JM. Skyrme-Hartree-Fock-Bogoliubov nuclear mass formulas: Crossing the 0.6 MeV accuracy threshold with microscopically deduced pairing. Physical Review Letters. 2009;**102**:152503

[21] Möller P, Nix JR, Myers WD, Swiatecki WJ. Nuclear ground-state masses and deformations. Atomic Data and Nuclear Data Tables. 1995;**59**: 185-381

[22] Wapstra AH, Audi G, Thibault C. The Ame2003 atomic mass evaluation: I. evaluation of input data, adjustment procedures. Nuclear Physics A. 2003; **729**:129-336

[23] Cline CK, Blann M. The pre-equilibrium statistical model: Description of the nuclear equilibration process and parameterization of the model. Nuclear Physics A. 1971;**172**: 225-229

[24] Dobes J, Betak E. Two-component exciton model. Zeitschrift für Physik. 1983;**310**:329-338

[25] Betak E, Dobes J. The finite depth of the nuclear potential well in the exciton model of preequilibrium decay. Zeitschrift für Physik A. 1976;**279**: 319-324

[26] Fu CY. Implementation of on advanced pairing correction for particle-hole state densities in precompound nuclear reaction theory. Nuclear Science and Engineering. 1984;**86**:344-354

[27] Kalbach C. Surface and collective effects in preequilibrium reactions. Physical Review C. 2000;**62**:44608

[28] Kalbach C. Surface effects in the exciton model of preequilibrium nuclear reactions. Physical Review C. 1985;**32**: 1157

Chapter 3

Some Key Issues of Vacuum System Design in Accelerators and Colliders

Jie Wang and Sheng Wang

Abstract

As we all know, vacuum system is the essential part for the accelerators and colliders, which provide the vacuum environment to minimize beam-gas interactions and maintain normal operation of the beams. With the proposals of future accelerators and colliders, such as Future Circular Collider (FCC), Super Proton-Proton Collider (SPPC), and International Linear Collider (ILC), it is time to review and focus on the key technologies involved in the optimization designs of the vacuum system of various kinds of accelerators and colliders. High vacuum gradient and electron cloud are the key issues for the vacuum system design of high-energy accelerators and colliders. This chapter gives a brief overview of these two key issues of vacuum system design and operations in high-energy, high-intensity, and high-luminosity accelerators and collider.

Keywords: vacuum system, beam-gas interaction, electron cloud, secondary electron yield, non-evaporable getters

1. Introduction

The high-energy accelerators and colliders, such as the Intersecting Storage Rings (ISR) [1, 2] at European Organization for Nuclear Research (CERN), the Super Proton Synchrotron (SPS) (CERN) [3, 4], the Tevatron proton-antiproton collider (United States) [5, 6], the abandoned Superconducting Super Collider (SSC) (United States) [7, 8], the Very Large Hadron Collider (VLHC) (United States) [9–11], the Large Hadron Collider (LHC) [12–14], the High-Luminosity Large Hadron Collider (HL-LHC) [15], and the High-Energy Large Hadron Collider (HE-LHC) [16, 17], have been proposed one after another for the discoveries and the establishing of standard model of particle physics.

Circular Electron Positron Collider (CEPC) (120 GeV) [18, 19] and the upgraded stage of Super Proton-Proton Collider (SPPC) (100 TeV) [20–22] based on lepton-proton colliders were proposed in China to explore the Higgs physics and the new physics beyond standard model, respectively. Moreover, in order to precisely study the flavour physics, such as the top particles, Higgs, Z and W, a luminosity-frontier, low-emittance and highest-energy electron-positron collider (FCC-ee) was proposed by the scientists from European Organization for Nuclear Research (CERN) [23, 24]. As the secondary stage, an energy-frontier hadron collider (FCC-hh) will be used to explore the possibility of existence of the dark matter candidates and get

Parameter	CEPC	FCC-hh	FCC-ee	HE-LHC	HL-LHC	LHC (pp)
E_{beam} [GeV]	120	50,000	45.5–175	12,500	7000	7000
Luminosity per IP [$10^{34}cm^{-2}s^{-1}$]	1.8	5–30	1.55–230	28	5	1
Circumference C [km]	54	97.8	97.75	26.7	26.7	26.7
Beam current [mA]	16.6	500	5.4–1390	1120	1120	584
SR power per beam [kW]	50	2400	50	100	3.6	0.0036

Table 1.
Key parameters for CEPC [29], FCC-hh [30], FCC-ee [26, 31], HE-LHC [30, 32], HL-LHC [30], and LHC(pp) [29].

the detailed information of Higgs self-coupling and the mechanism of electroweak symmetry breaking. FCC-ee [25, 26] and FCC-hh [27] are the two stages of Future Circular Collider (FCC) [28]. FCC has a center-of-mass energy of 100 TeV with proton-proton collisions finally. The key parameters of CEPC, FCC-hh, FCC-ee, HE-LHC, HL-LHC, and LHC are shown in **Table 1**.

Beam-related instabilities and electron cloud are the critical aspects for vacuum system of the high-energy, high-intensity, and high-luminosity accelerators, which could affect the machine performance and operation [33].

2. Electron cloud

Electron cloud (EC) issue is one of the important aspects for high-energy accelerators and colliders. The primary electrons produced by the ionization of residual gases or by photoemission are accelerated by the beam and impact on the vacuum chambers and generate secondary electrons. Then, the secondary electrons can be reflected, accelerated, or absorbed in the vacuum chambers, even induce electron avalanche.

Secondary electron yield (SEY) is an important parameter to understand the formation and dissipation of the EC in accelerators [34]. The average of SEY over all electron-wall collisions for a certain time is the effective SEY (δ_{eff}), which depends on the chamber and beam parameters. When δ_{eff} is less than 1, the net number of secondary electrons balances that of electrons absorbed by inner surfaces of vacuum chambers. When δ_{eff} is larger than 1, the secondary electrons grow exponentially and then the EC reaches a dynamical equilibrium with δ_{eff} value equating to 1.

The electron accumulation can induce beam losses, beam instabilities, the emittance growth, the heating of the vacuum chambers, vacuum instabilities, and the decrease of detection precision [33], as shown in **Figure 1**. The detrimental effects of EC on beam quality have been observed and studied in many accelerators. The estimation and simulation of EC formation process and the exploration of EC inhibition methods are critical for the understanding of EC build-up and related effects.

2.1 Numerical simulations of electron cloud

High-energy beam-induced synchrotron radiation can result in the increase of vacuum pressure and the primary photoelectrons. These can finally contribute to the EC formation. Many parameters can affect the EC formation, such as beam energy [35], bunch spacing [36], bunch size, bunch intensity, vacuum pressure, the geometry of the vacuum pipes [37], the properties of inner surface of vacuum chamber [38], the secondary electron energy spectrum [34], etc.

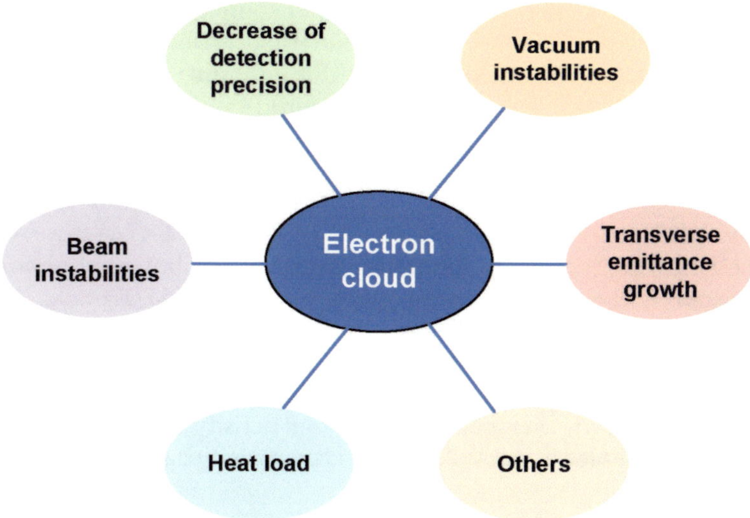

Figure 1.
Electron cloud-induced related effects.

There are three kinds of EC simulation codes: build-up simulation codes, instability simulation codes, and self-consistent simulation codes. The EC density distribution as the function of time and energy and the details of the interaction between electrons and walls can be obtained based on the EC build-up simulation codes with 2D and 3D versions. For the study of EC in isolated regions like field-free regions, 2D codes are preferable. For the study of EC in magnetic regions like fringe fields regions, 3D codes are usually adopted. EC build-up simulation codes cannot get the information on the interaction between the EC and the beam [39], while instability simulation codes can get the details on the interaction between dynamical beam particles and the prescribed EC. For instability simulation codes, the interaction between electrons and the walls is simplified or negligible. Self-consistent simulation codes are more computational than the above two simulation codes, to better understand the dynamic effects between the electrons and the beams, such as WARP-POSINST codes [40].

Guillermo et al. studied the details of the build-up of electron cloud and the effect of synchrotron radiation on EC formation based on the code of Synrad3D and the 2D macro-particle code of PyECLOUD [37]. The results demonstrated that the different surface conditions could influence the heat load and average electron line density. The smoother the surfaces, the higher the average linear density for all SEY values selected. For smooth surface, the heat load is the highest. Moreover, the heat load on the inverted sawtooth vacuum surface was slightly smaller than that on the oriented sawtooth surfaces.

The EC build-up and the effect of EC on the proton beam in the SPS accelerator were simulated by Vay et al. [40], based on the WARP-POSINST codes. The analysis results indicated that the interaction between EC and the proton bunches can induce the increase of emittance, vertical bunch size, leading to the increase of electron density.

The Fortran code ECLOUD was developed to understand the EC effect depending on the specified secondary emission model [41, 42]. The true secondary electrons and reflected electrons and their energy distribution are included in this code, which contribute to the secondary emission. Various shapes such as round pipes, elliptical pipes, and rectangular pipes can be simulated in this code. The detector effect and the scrubbing effect can be simulated to better compare with the experimental results. The general layout of the ECLOUD code (first version in 1997 at CERN by Zimmermann) [43], the PEI code (in 1995 at KEK by Ohmi) [44], and the POSINST code (in 1996 at LBNL by Furman and Lambertson) [45] is basically similar.

The POSINST code was employed by Crittenden et al., to study the EC build-up in the arc dipole region of positron damping ring of International Linear Collider (ILC) [46]. The secondary electron emission (SEE) parameters of a TiN surface tested at Cornell Electron Storage Ring Test Accelerator (CESRTA) were used in the SEY model during the POSINST simulations. The distribution of photons and photon transport in the arc region were calculated by SYNRAD3D code. For the estimations of EC densities in quadrupoles, sextupoles, and field-free regions, the ECLOUD code [41, 47] was used. The simulation results demonstrated that the beam emittance growth during 300 turns and 18,550 turns were 0.16 and 10% because of ECs. When the averaged electron cloud density was 3.5×10^{10} m^{-3}, the beam emittance increased about 0.2% as the chromaticity values increased from 0 to 6 [46].

The 3D code CLOUDLAND was adopted by Wang et al., to study the EC evolution in the quadrupole and sextupole regions of the ILC/CESRTA with and without ante-chamber [48]. The simulation results indicated that the average EC density in the quadrupole magnet of the ILC can be reduced by 98% under the presence of ante-chamber when the SEY was less than 1.1. When the SEY was larger than 1.1, the effect of ante-chambers on EC densities was not notable. Larger SEY can induce higher EC average densities. The strong space charge can reduce the EC density and change the EC distribution dramatically. Therefore, the reduction of SEYs in quadrupole and sextupole regions is important for the decrease of the photon flux.

2.2 Single and coupled bunch instabilities

When the head of the bunch is diverged from the beam axis and interacts with the EC, it will induce the single bunch head-tail instability [49]. The following bunches will be affected by the new EC distributions and the tails can be deflected finally. The property of EC here is similar to the short-range wake field. The electron distribution, the energy spectrum of electrons, and the evolution of the EC can be simulated and compared with the experimental results. Then, the EC densities were obtained and used for the input parameters to analyze the impact of EC on the bunches.

The bunch instability caused by EC in a linac can induce the beam break-up in the case without synchrotron radiation. And in the case of synchrotron radiation, the instability caused by EC is analogous to the transverse mode coupling instability (TMCI) [50]. The impacts of space charge, magnetic field, the chromaticity, the amplitude detuning, and the broad-band resonator are also considered in the study of the bunch instability caused by EC [50–53].

The single and coupled bunch instabilities are mainly related to the frequency of the EC, the synchrotron tune, the machine circumference, the beam sizes, the bunch length, the chromaticity, and the relativistic factor [54], as shown in **Figure 2**.

Various single bunch instability codes such as MICROMAP (developed at GSI) [55, 56], PEHTS (developed at KEK) [35, 57–60], CMAD [61, 62], and HEADTAIL (developed at CERN) [52, 63] were developed to study the related effects.

The CMAD code, the HEADTAIL code, and the WARP code were used by Li et al. [62] to study the effects of various wideband feedbacks on the EC instabilities in CERN SPS, which provided valuable information for HL-LHC. The results showed that a bandwidth of 500 MHz was needed to mitigate the EC when the EC

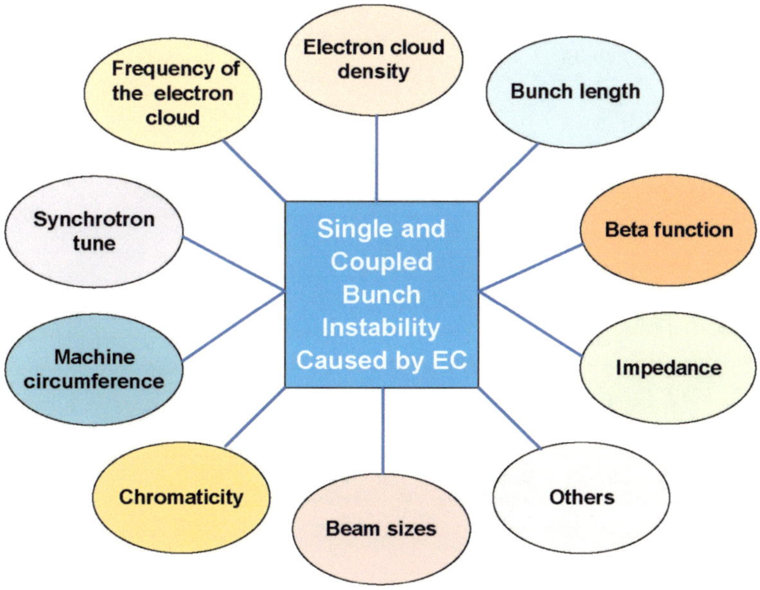

Figure 2.
Key parameters for the single and coupled bunch instability induced by electron cloud.

densities were above 1×10^{12} m^{-3}. When the EC densities were above 1.4×10^{12} m^{-3}, a bandwidth of 1 GHz may be required.

The single bunch instability caused by EC with different densities in LHC and SPS was investigated by E. Benedetto et al. [52] based on the HEADTAIL simulation code [53, 63]. The EC effect on transverse single bunch instabilities in the bending region was simulated and compared with observations in the SPS. And for the LHC, the simulation results indicated that the chromaticity is a key factor for the head-tail instability, but may not for a short- and long-term emittance growth [52].

Based on the HEADTAIL code and ECLOUD program, the simulation results demonstrated that the instability growth process and the chromaticity beneficial effect were consistent with the observed results, which were reported by G. Rumolo et al. [42].

Ohmi studied the interaction between the EC and the single bunch via PIC method (PEHTS) in High Energy Accelerator Research Organization B-factory Low Energy Ring (KEKB LER) [57]. The simulation results showed that the strong head-tail instability in KEKB-LER was related to the coherent instability caused by the EC.

In Beijing Electron Positron Collider (BEPC), the single and coupled bunch instabilities caused by EC were studied systematically by Wang et al. [51]. The effects of ante-chamber, TiN films, and clearing electrodes on electron cloud instability (ECI) were explored by simulations and experiments. The simulation and experimental results manifested that the single bunch instability can be reduced below the threshold and the coupled bunch instability can be restricted by the feedback systems.

2.3 Electron cloud mitigation methods

Various methods, such as the electrode cleaning, the solenoid, the beam scrubbing, the film coatings, and geometrical modification (like laser processed surfaces), have been developed for EC mitigation in accelerators and colliders, as shown in **Figure 3** [64–67]. The test results in Relativistic Heavy Ion Collider (RHIC) showed that the solenoids with the magnetic field of 0.5 mT were effective for the EC suppression [64].

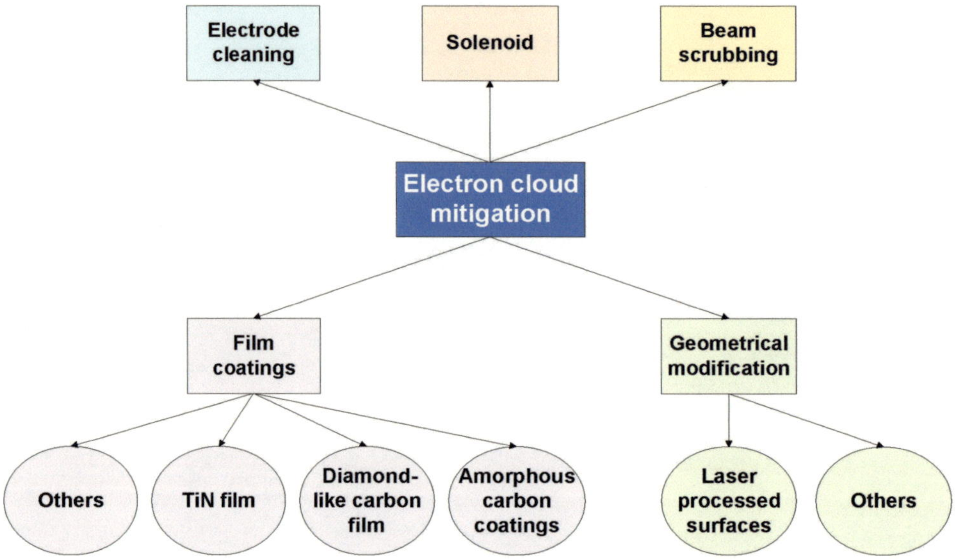

Figure 3.
Electron cloud mitigation methods.

With the advantage of no additional parts introduced, film coatings (like TiN coatings, diamond-like carbon film, highly oriented pyrolytic graphite, and the amorphous carbon film) [66] and geometrical modification methods were studied widely.

Compared to other films, TiN coatings were applied in accelerators and colliders extensively. Therefore, the related properties and the application of TiN film will be introduced emphatically.

Recently, laser ablation method was proposed as a novel and potential way for reducing the SEY [65, 68] with several advantages, such as low cost, no requirement of vacuum, etc. The impedance issue, desorption properties, and the compatibility with cryogenic vacuum should be evaluated and studied carefully in the future accelerators and colliders.

2.3.1 TiN film

In order to mitigate EC in accelerators, TiN film coatings were employed in the inner surface of vacuum chambers [69–75]. The experimental results indicate that TiN film coating is an effective way for EC mitigation. The effects of substrate materials, the stoichiometry, film thicknesses, ion/electron bombardments, and deposition parameters on the SEY and related properties of TiN have been investigated systematically [70, 72, 74, 76–83].

TiN film coatings were deposited on the inner surfaces of copper beam ducts in KEK by Shibata et al. [74, 75]. As an example, a pipe length of 3.6 m was coated and installed in the KEK B-factory positron ring using a titanium rod of 4.2 m long. The maximum SEY (δ_{max}) of the TiN film coatings varied between 0.70 and 1.58 with the electron doses ranging between 10^{-1} and 10^{-5} C mm^{-2}. The coated chambers were tested in LER. The test results manifested that the EC density in the coated chambers decreased about ~33% comparing to that of the uncoated ones at the beam current of 800 mA.

In Laboratory of the Linear Accelerator-Orsay (LAL-Orsay), TiN film coating was used as the EC inhibition method in the RF ceramic windows [73]. The planar and cylindrical alumina ceramics were the substrates with the properties of high mechanical strength, high stability, and low outgassing rate. The SEY of the alumina ceramic was decreased by depositing the TiN film on its surfaces, which will benefit for the avoiding of electron multipactor.

The effects of H_2 gas, ion, and electron conditioning on the SEY of TiN coating for EC suppression in storage rings were investigated by F. Le Pimpec et al. in Stanford Linear Accelerator Center (SLAC) [78] and Lawrence Berkeley National Laboratory (LBNL) [70, 72, 84]. The δ_{max} of TiN coating was reduced from 1.50 to 1.10 after N^{2+} ion bombardment. During N^{2+} ion bombardment, the contaminants, such as hydrocarbons, water, and oxide with high SEY, were removed, which contributed to the SEY decrease of TiN coating. Ion bombardment could induce the interstitial N, while the vacancies may be filled by the nitrogen from the beam pipe [85]. The δ_{max} of TiN coating with Al substrates ranged from 1.52 to 1.99. The thickest TiN coatings with Al or stainless steel substrates had the lowest SEY here. As the electron conditioning doses increased, the maximum energy (E_{max}) corresponding to the δ_{max} shifted to lower energy and the SEY curves bend disappeared. This SEY evolution may be related to the removal or dissociation (from TiO_2 into defective suboxide) of the surface contaminants. When the TiN coatings were baked out in vacuum at 150°C for 2 h, the δ_{max} decreased from 1.70 to 1.60 [70]. For the TiN coating deposited on the grooved Al surface, the δ_{max} of TiN coatings decreased to 1.30 from 2.30.

The vacuum pipes of different sizes in Spallation Neutron Source (SNS) were coated with TiN films by reactive DC magnetron sputtering in Brookhaven National Laboratory (BNL) [64, 71]. The arc regions, straight sections, injection kicker ceramic chambers, and extraction kicker modules were deposited with TiN, with the lengths of 0.5–5 m and the diameters of 20–36 cm.

The TiN films were coated on the racetrack-type ceramic pipe for the EC suppression in National Synchrotron Radiation Laboratory by Wang et al. [86, 87]. CST PARTICLE STUDIO software was used to optimize the TiN film thickness in magnetron sputtering system. Moreover, two kinds of Ti cathodes (Ti rods and Ti plates) were adopted and compared to improve the TiN film deposition rate.

In order to reduce the electron multipactor effects and improve power transmission in the TESLA couplers at DESY, TiN films were applied on the inner surface of the wave guides and the RF windows [88]. The performance of low SEY remained steady even after 24 h air exposure. In addition, the RF conditioning time reduced from ~3 days to 4–6 h due to the TiN films.

Above all, the thicknesses, SEYs, and substrates of the TiN films were summarized in **Table 2**. The film thicknesses of TiN films were usually ranging from 7 to 200 nm. Aluminum alloy, stainless steel, copper, and ceramic are generally chosen as the substrates.

2.3.2 Other films

Besides TiN coatings, other low SEY film coatings, like the carbon film [66], highly oriented pyrolytic graphite (HOPG) [66], the diamond-like carbon film (DLC) [89], and the amorphous carbon coatings (a-C) [90] will be introduced in this section.

Ceramic pipes were used in the rapid-cycling synchrotron (RCS) of the Japan Proton Accelerator Research Complex (J-PARC) to avoid the eddy current effect [89]. In order to reduce the electron emission from the internal surfaces of the ceramic pipes, the DLC coating was applied as a possible alternative method. The experimental results indicated that the SEYs of DLC coating exposing to the water/oxygen/electron beam were 1.15–1.20 and less than that of TiN coatings (1.50–1.70).

Institutions	Substrates	Thickness/nm	δ_{max}
KEK [74, 75]	Copper (beam ducts)	100–200	1.05–1.60
LAL [73]	Ceramic (RF windows)	7–15	1.50
SLAC & LBNL [72, 78, 84]	6063 aluminum alloy/ stainless steel	100–204	1.01–1.99 (ion and electron conditioning)
BNL [64, 71]	Ceramic/stainless steel (chambers)	~100	Less than 1.80
DESY [88]	Ceramic/aluminum/ copper (coupler windows, couplers)	4–40	/
NSRL [86, 87]	Ceramic	45–3200	/

Table 2.
The SEYs, thicknesses, substrates of TiN film coatings prepared by various research institutions. Here, δ_{max} is the maximum SEY within the primary electron energy considered.

Amorphous carbon films with the δ_{max} of 0.92–1.33 have been used on the vacuum pipes for the EC elimination in SPS with LHC-type beam [90–92]. The test results demonstrated that the EC can be suppressed in the liners even after air exposure for 3 months. Moreover, the properties of a-C films were stable after 3 years beam operations.

Carbon films [93] were prepared with the δ_{max} of ~0.97, which was lower than that of HOPG (~1.23) [66]. The SEY difference between carbon films and HOPG can be ascribed to that the density of carbon films was smaller and the larger penetration range and the SEs scattering to defects/pores than that of HOPG. The XPS results showed that the SEY values were related to the C 1 s high binding energy. When the concentration of oxygen was below 16 at%, it would not affect the SEY of carbon films.

3. High vacuum gradient

The rough pumps, the turbo molecular pumps, and the ion pumps are usually adopted between the ends of vacuum pipes in accelerators and colliders. These pumps can reduce the pressure efficiently near it, while the pressure in the middle of the vacuum pipes is higher than the ones near the pumps. In other words, the application of traditional pumps can induce the high vacuum gradient. In order to reduce the vacuum gradient in the long vacuum pipes, non-evaporable getter (NEG) coatings were proposed to solve this issue [94–99].

3.1 NEG

Generally, NEGs are the alloys Fe, Al, Zr, etc., with the property of absorbing residual gases, like H_2, CO, and CH_4, after high-temperature activation in vacuum systems. NEGs are widely used in accelerators and colliders for the achievement of ultra-high vacuum (UHV) and provide distributed pumping [97, 100, 101]. The H_2 outgassing rate for aluminum alloys or stainless steel is about 10^{-13} Torr l s^{-1} cm^{-2}, which is the main obstacle for UHV achieving. In order to reduce the outgassing rate and achieve the required ultimate pressure, NEG alloys were applied in vacuum systems.

3.1.1 NEG binary NEG alloy

Binary NEG alloy Zr-Al (St101) [102] was prepared on both sides of strips of getter pumps with the activation temperature of ~700°C and applied in the vacuum pipes of Large Electron Positron (LEP) Collider at CERN.

Zr-Fe (St198) NEG alloy has been applied in various scenarios, such as the N_2-filled devices, the integrated electronic circuit for N_2 purifier, the conversion of tritiated water, hydrogen storage devices, and the hydrogen isotopes absorption and desorption [103]. The XPS results indicated the surface chemical state changes during thermal activation. At higher activation temperatures, the concentration of metallic Zr improved in the surfaces. The concentration of carbon increased gradually under the case of activation temperature of less than 400°C and then decreased.

The ESD, pumping speed, and ultimate pressure of Ti-Hf, Hf-Zr, and Ti-Zr binary [98] NEG were studied and compared by C. Benvenuti et al. Thereinto, the Ti-Zr NEG alloy has the lowest activation temperature of ~200°C. The ultimate pressure of ~10^{-11} Pa can be obtained using the binary coatings, which can provide the pumping speed for H_2 of ~$0.5\,l\,s^{-1}\,cm^{-2}$.

3.1.2 Ternary NEG alloy

Ternary NEG alloys, like Ti-Zr-V and Zr-V-Fe (St707, St737) NEG films [96, 104–107], have been applied in the vacuum system of accelerators and colliders. Ti-Zr-V NEG coatings have the activation temperature of less than 200°C [108]. The absorption and desorption mechanisms during thermal activation and pumping properties of ternary NEG alloy coatings have been investigated extensively.

The research results showed that the concentrations of Ti and Zr were enriched on the surface of activated Ti-Zr-V NEG films [108, 109]. The oxygen detected on the activated surface was mainly in the form of Zirconium suboxides and titanium suboxides.

For the sake of reducing the activation temperature of ternary NEG alloys, different compositions of Ti, Zr, and V elements were produced on stainless steel surfaces. The surface chemical state and crystal structure variations of Ti, Zr, and V elements were analyzed by Auger electron spectroscopy (AES) and X-ray diffraction (XRD) during thermal activation [94]. The getters with a nanocrystalline structures could be activated more easily as the activation behavior is related to the solubility and the diffusion of oxygen.

3.1.3 Quaternary NEG alloy

Quaternary Ti-Zr-Hf-V NEG films were proposed by Malyshev with the advantage of lowest activation temperature of ~150°C [110–112]. Comparing to the ESD yields for all desorbed species (H_2, CO, and CO_2) of Ti-Zr-V NEG films, those of Ti-Zr-Hf-V NEG films were lower. Two kinds of Ti-Zr-Hf-V NEG films, the dense one and the columnar one, were prepared. The experimental results demonstrated that the columnar one has lower initial ESD yields, the higher pumping speeds, and capacities for all desorbed species than that of the dense one. In addition, a dual-layer NEG coating was prepared with the bottom dense layer and the top columnar layer. Here, hydrogen diffusion can be barred by the bottom layer. The ESD of hydrogen can be improved and the pumping properties can be enhanced by the top layer.

To understand the effect of the coatings on the wakefield impedance, the surface resistance of Ti-Zr-Hf-V NEG films was tested and analyzed at 7.8 GHz [110]. Based on the analytical model, the conductivities of dense and columnar NEG coatings were 8×10^5 S/m and 1.4×10^4 S/m. The surface resistance of copper increased slightly after NEG coating depositions.

3.2 Photons-/electrons-/ions-stimulated desorption

The photons-/electrons-/ions-induced gas desorption may result in pressure instability and then lead to the beam loss. The ionization of residual gases is proportional to the beam current. Therefore, the materials, the geometry, and the pumping speed of the vacuum systems should be considered carefully [113–120]. In order to reach the base pressures of $\sim 10^{-9}$ Pa and reduce the pressure instability, low photon-stimulated gas desorption (PSD) yield, electron-stimulated gas desorption (ESD) yield, and ion-stimulated gas desorption (ISD) yield materials are preferable in vacuum systems.

The ESD and ISD yield of the commonly used materials like the OFHC, the stainless steel in accelerators and colliders, were measured and analyzed during the bake out at 150–600°C [114]. The test results indicated that the surfaces of metallic oxide may have the porous structures, which can trap the residual gases such as H_2 and CH_4.

The ion-induced pressure instability, firstly observed at CERN, can limit the beam current in accelerators and colliders and has been studied intensively [121–123]. The effects of the ion dose, the mass, and the energy on the ISD yields of copper and aluminum were investigated [119]. The desorption yields of the copper and the aluminum decreased by two times after the bake out. NEG coatings are the mostly used solution for reducing the ion-induced vacuum stability [121].

It was found that the application of beam screen can effectively reduce the vacuum instability caused by ISD, ESD, and PSD at room temperature and cryogenic regions in LHC [118]. For example, the average H_2 density caused by PSD could be reduced by over 50 times by incorporating the beam screen in the vacuum pipes [124]. Moreover, glow discharge and baking are also useful methods for the reduction of ion-induced pressure instability [125, 126].

4. Final remarks

This chapter introduces the two key issues: electron clouds and high vacuum gradient. Electron cloud issue may influence the beam quality and stability in accelerators and colliders. Various methods, such as the DLC film, the a-C film and laser processed surfaces, have been proposed for the EC mitigation in the warm and cryogenic regions. As for the high vacuum gradient issue, the preparation and properties of binary/ternary/quaternary NEG alloy films have been studied to decrease the activation temperature and beam-gas interactions and also to improve the pumping properties, PSD, ESD, and ISD yields at room temperature and cryogenic temperature. For the design of UHV vacuum system of accelerators and colliders, the quality, the conformity, and the engineering aspects should be investigated and analyzed carefully. In regard to vacuum operations, the safety, reliability, and the machine performance limitations should be considered and tested carefully.

Acknowledgements

We would like to thank our colleagues from Xi'an Jiaotong University: Prof. Shaoqiang Guo, Prof. Zhanglian Xu, and Ms. Jing Zhang for useful suggestions.

This research was funded by the National Natural Science Foundation for the Youth of China No. 11905170, the Fundamental Research Funds for the Central Universities No. XJH012019018, and the National Natural Science Foundation of China under Grant No. 11775166.

Conflict of interest

The authors declare no conflict of interest.

Author details

Jie Wang and Sheng Wang*
Shaanxi Key Laboratory of Advanced Nuclear Energy and Technology, Shaanxi Engineering Research Centre of Advanced Nuclear Energy, School of Nuclear Science and Technology, School of Energy and Power Engineering, Xi'an Jiaotong University, Xi'an, Shaanxi, China

*Address all correspondence to: shengwang@xjtu.edu.cn

References

[1] Johnsen K. CERN intersecting storage rings (ISR). Proceedings of the National Academy of Sciences of the United States of America. 1973;**70**(2):619-626. DOI: 10.1073/pnas.70.2.619

[2] Carboni G, Owen DL, Ambrosio M, Anzivino G, Barbarino G, Paternoster G, et al. Evidence of a rise in the antiproton-proton total cross section at the CERN intersecting storage rings. Physics Letters B. 1982;**113**(1):87-92. DOI: 10.1016/0370-2693(82)90115-0

[3] Evans L. The proton-antiproton collider. In: Third John Adams Memorial Lecture. Geneva: CERN; 1988;**88**:01. DOI: 10.5170/CERN-1988-001

[4] van Hees H, Rapp R. Dilepton radiation at the CERN super-proton synchrotron. Nuclear Physics A. 2008;**806**(1-4):339-387. DOI: 10.1016/j.nuclphysa.2008.03.009

[5] Acosta D, Affolder T, Akimoto T, Albrow MG, Ambrose D, Amerio S, et al. First measurements of inclusive W and Z cross sections from run II of the Fermilab Tevatron collider. Physical Review Letters. 2005;**94**:091803. DOI: 10.1103/PhysRevLett.94.091803

[6] Holmes S, Moore RS, Shiltsev V. Overview of the Tevatron collider complex: Goals, operations and performance. Journal of Instrumentation. 2011;**6**(08):T08001-T. DOI: 10.1088/1748-0221/6/08/T08001

[7] Jackson JD, Barton RG, Donaldson R, Savage DK. Conceptual Design of the Superconducting Super Collider [Internet]. 1986. Available from: http://inspirehep.net/record/229226/files/ssc-sr-2020.pdf [Accessed: 11 March 2020]

[8] Kevles DJ. Big Science and Big Politics in the United States: Reflections on the Death of the SSC and the Life of the Human Genome Project. United States: University of California Press; 1997. DOI: 10.2307/27757780

[9] Group TVds. Design Study for a Staged Vary Large Hadron Collider. Stanford, CA: SLAC; 2001;R:591. DOI: 10.2172/781994

[10] Burov A, Marriner J, Shiltsev V, Danilov V, Lambertson G. Beam stability issues in very large hadron collider. Nuclear Instruments and Methods in Physics Research A. 2000;**450**:194-206. DOI: 10.2172/781994

[11] Sen T, Norem J. Very large lepton collider in the very large hadron collider tunnel. Physical Review Special Topics - Accelerators and Beams. 2002;**5**(3):031001. DOI: 10.1103/PhysRevSTAB.5.031001

[12] Benda V, Bézaguet A, Casas-Cubillos J, Claudet S, Erdt W, Lebrun P, et al. Conceptual Design of the Cryogenic System for the Large Hadron Collider (LHC). European Organization for Nuclear Research: Sitges, Barcelona; 1996. DOI: 10.1134/S106377881012104X

[13] Evans L. The large hadron collider. New Journal of Physics. 2007;**9**(19):335. DOI: 10.1098/rsta.2011.0453

[14] Potter KM. The large hadron collider (LHC) project of CERN. European Organization for Nuclear Research. LHC-Project-Report-36. 1996:1760-1763. DOI: 10.1134/S106377881012104X

[15] Apollinari G, Brüning O, Nakamoto T, Rossi L. High luminosity Large Hadron Collider HL-LHC. United States; 2015. Report No.: CERN-2015-005; FERMILAB-DESIGN-2015-02-1418897; TRN: US1701927. DOI: 10.5170/CERN-2015-005.1

[16] Zimmermann F. HE-LHC Overview. Parameters and challenges. 2017;**72**:138-141. Available from: http://cds.cern.ch/record/2315725/files/9999999_138-141.pdf

[17] Keintzel J, Crouch M, Hofer M, Risselada T, Tomás R, Zimmermann F. HE-LHC optics design options. In: 10th Int Partile Accelerator Conf. Melbourne, Australia: JACoW Publishing; 2019. pp. 492-495

[18] Group TCS. CEPC conceptual design report volume I – Accelerator; 2018

[19] Gao J. CEPC-SPPC towards CDR. In: Proceedings of IPAC2017. Copenhagen, Denmark: JACoW Publishing; 2017. pp. 2954-2957

[20] Canbay AC, Kaya U, Ketenoglu B, Oner BB, Sultansoy S. SppC based energy frontier lepton-proton colliders: Luminosity and physics. Adv. High Energy Phys. 2017;**2017**:1-6. DOI: 10.1155/2017/4021493

[21] Wang L, Tang J, Ohmi K. Beam-beam studies for super proton-proton collider. In: 9th International Particle Accelerator Conference. Vancouver, BC, Canada: JACoW Publishing; 2018. pp. 2918-2920

[22] Su F, Gao J, Yukai Chen JT, Wang D, Wang Y, Bai S, et al. SPPC parameter choice and lattice design. In: Proceedings of the 7th International Particle Accelerator Conference. Busan, Korea, Switzerland: JACoW Publishing; 2016. pp. 1400-1402

[23] Garion C, Kersevan R. Design of the vacuum system of the FCC-ee electron-positron collider. In: 10th International Particle Accelerator Conference (IPAC2019). Melbourne, Australia: JACoW Publishing; 2019. pp. 1319-1322

[24] Lesiak T. Flavour physics at the FCC-ee. Acta Physica Polonica B. 2018;**49**(6):1241-1246. DOI: 10.5506/APhysPolB.49.1241

[25] Koratzinos M. FCC-ee accelerator parameters, performance and limitations. Nuclear and Particle Physics Proceedings. 2016; **273-275**:2326-2328. DOI: 10.1016/j.nuclphysbps.2015.09.380

[26] Benedikt M, Oide K, Zimmermann F, Bogomyagkov A, Levichev E, Migliorati M, et al. Status and challenges for FCC-ee [Internet]. 2015. Report No.: CERN-ACC-2015-0111. Available from: https://arxiv.xilesou.top/ftp/arxiv/papers/1508/1508.03363.pdf

[27] Abada A, Abbrescia M, AbdusSalam SS, Abdyukhanov I, Abelleira Fernandez J, Abramov A, et al. FCC-hh: The hadron collider. The European Physical Journal Special Topics. 2019;**228**(4):755-1107. DOI: 10.1140/epjst/e2019-900087-0

[28] Naseem S, Nasri S, Soualah R. Dark matter searches as new physics at the future circular collider (FCC). Journal of Physics: Conference Series. 2019;**1258**:1258. DOI: 10.1088/1742-6596/1258/1/012017

[29] Zimmermann F, Benedikt M, Schulte D, Wenninger J. Challenges for highest energy circular colliders. In: Proceedings of 5th International Particle Accelerator Conference; 15-20 Jun 2014. Dresden, Germany, Geneva: JACoW; 2014. p. 7. DOI: 10.1103/PhysRevD.70.033011

[30] Benedikt M, Zimmermann F. FCC colliders at the energy frontier. In: 9th International Particle Accelerator Conference. Vancouver, BC, Canada: JACoW Publishing; 2018. pp. 2908-2913

[31] Belli E, Castorina G, Migliorati M, Persichelli S, Rumolo G, Spataro B, et al. Some critical collective effects for the FCC-ee collider. ICFA Beam

Dynamics Newsletters. 2017;**72**:61-69. Available from: http://cds.cern.ch/record/2299210/files/belli.pdf

[32] Bartmann W, Benedikt M, Besana MI, Bruce R, Brüning O, Buffat X, et al. Beam Dynamics Issues in the FCC. Geneva: JACoW; 2016. pp. WEAM5X01. DOI: 10.18429/JACoW-HB2016-WEAM5X01

[33] Belli E, Castorina G, Migliorati M, Persichelli S, Rumolo G, Spataro B, et al. Some Critical Collective Effects for the FCC-ee Collider [Internet]. 2017. Available from: http://cds.cern.ch/record/2299210/files/belli.pdf [Accessed: 11 March 2020]

[34] Furman MA. Electron cloud effects in accelerators. arXiv:13101706v1 [physicsacc-ph]. 2013. DOI: 10.5170/CERN-2013-002.8

[35] Rumolo G, Arduini G, Metral E, Shaposhnikova E, Benedetto E, Calaga R, et al. Dependence of the electron-cloud instability on the beam energy. Physical Review Letters. 2008;**100**(14):144801. DOI: 10.1103/PhysRevLett.100.144801

[36] Furman MA. Studies of e-cloud build up for the fnal main injector and for the LHC. In: 397th ICFA Advanced Beam Dynamics Workshop - Highintensity High Brightness Hadron Beams; Tsukuba, Japan. United States; 2006

[37] Guillermo G, Cuna GHIM, Ortiz EDO, Zimmermann F. Electron cloud build up for LHC 'sawtooth' vacuum chamber. In: 9th International Particle Accelerator Conference. Vancouver, BC, Canada: JACoW Publishing; 2018. pp. 744-746

[38] Cimino R, Collins IR, Furman MA, Pivi M, Ruggiero F, Rumolo G, et al. Can low-energy electrons affect high-energy physics accelerators? Physical Review Letters. 2004;**93**(1):014801. DOI: 10.1103/physrevlett.93.014801

[39] Chin YH, editor. KEK Proceedings 96-6, Proceedings of the International Workshop on Collective Effects and Impedance for B-Factories. Tsukuba, Japan; 1995

[40] Vay J, Furman M, Venturini M. Direct numerical modeling of E-cloud driven instability of three consecutive batches in the CERN SPS. In: Proceedings of International Particle Accelerator Conference IPAC'12, 20-25 May 2012. New Orleans, Louisiana. USA; 2012. pp. 1125-1127

[41] Schulte D, Zimmermann F. Electron Cloud Build-up Simulations Using ECLOUD. Geneva: CERN; 2004:143-152. DOI: 10.5170/CERN-2005-001.143

[42] Rumolo G, Zimmermann F. Electron cloud simulations: Beam instabilities and wakefields. Physical Review Special Topics - Accelerators and Beams. 2002;**5**(12):121002. DOI: 10.1103/PhysRevSTAB. 5.121002

[43] Zimmermann F. A simulation study of electron-cloud instability and beam-induced multipacting in the LHC. In: CERN CH-1211. Geneva, Switzerland: CERN; 1997. DOI: 10.2172/291055

[44] Ohmi K. Beam-photoelectron interactions in positron storage rings. Physical Review Letters. 1995;**75**(8):1526-1529. DOI: 10.1103/PhysRevLett.75.1526

[45] Furman MA, Lambertson GR. The Electron-Cloud Instability in PEP-II. New York: IEEE; 1997:1617-1619. DOI: 10.1109/PAC.2001.987222

[46] Crittenden JA, Conway J, Dugan GF, Palmer MA, Rubin DL, Shanks J, et al. Investigation into electron cloud effects in the international linear collider positron damping ring. Physical Review Special Topics - Accelerators and Beams. 2014;**17**(3):031002. DOI: 10.1103/PhysRevSTAB.17.031002

[47] Rumolo G, Zimmermann F. Electron-cloud simulations build up and related effects. In: Mini Workshop on Electron Cloud Simulations for Proton and Positron Beams (Ecloud'02); 15-18 Apr 2002. Geneva, Switzerland; 2002. pp. 97-111. DOI: 10.5170/CERN-2002-001.97

[48] Wang L, Pivi M. Trapping of electron cloud in ILC-CESRTA quadrupole and sextupole magnets [Internet]. 2011. Available from: http://inspirehep.net/record/925671/files/slac-pub-14388.pdf [Accessed: 04 February 2020]

[49] Rumolo G, Zimmermann F. Theory and simulation of the electron cloud instability [Internet]. 2001. Available from: http://citeseerx.ist.psu.edu/viewdoc/download?doi=10.1.1.7.3749&rep=rep1&type=pdf [Accessed: 04 February 2020]

[50] Ohmi K, Zimmermann F, Perevedentsev E. Wake-field and fast head-tail instability caused by an electron cloud. Physical Review. E, Statistical, Nonlinear, and Soft Matter Physics. 2002;**65**(1 Pt 2):016502. DOI: 10.1103/PhysRevE.65.016502

[51] Wang JQ, Guo ZY, Liu YD, Qin Q, Xing J, Zhao Z. Electron cloud instability studies in the Beijing electron positron collider. Physical Review Special Topics - Accelerators and Beams. 2004;7:094401. DOI: 10.1103/PhysRevSTAB. 7.094401

[52] Benedetto E, Schulte D, Zimmermann F, Rumolo G. Simulation of transverse single bunch instabilities and emittance growth caused by electron cloud in LHC and SPS [Internet]. 2005. Available from: https://cds.cern.ch/record/847792/files/p331.pdf [Accessed: 04 February 2005]

[53] Rumolo G, Zimmermann F. Simulation of single bunch instabilities driven by electron cloud in the SPS. In: Proceedings of the 2001 Particle Accelerator Conference. Chicago; 2001. pp. 1886-1888. DOI: 10.1109/PAC.2001.987216 [Accessed: 04 February 2020]

[54] Belli E, Migliorati M, Rumolo G. Electron cloud and collective effects in the interaction region of FCC-ee. In: Proceedings of eeFACT. Daresbury, UK: JACoW; 2016. DOI: 10.18429/JACoW-eeFACT2016-TUT3AH7

[55] Benedetto E, Zimmermann F, Franchetti G, Ohmi K. Beam loss, emittance growth and halo formation due to the pinched electron cloud. In: Proceedings of HB. Tsukuba, Japan: JACoW; 2006

[56] Rumolo G. Electron cloud effects for PS2, SPS(+) and LHC. In: Proceedings HHH-2008. Geneva: CERN; 2009:115-119. DOI: 10.5170/CERN-2009-004.115

[57] Ohmi K. Particle-in-cell simulation of beam-electron cloud interactions. In: Proceedings of the Particle Accelerator Conference. Chicago: USA; 2001

[58] Benedetto E, Ruggiero F, Schulte D, Zimmermann F, Blaskiewicz M, Wang L, et al. Review and Comparison of Simulation Codes Modeling Electron-Cloud Build Up and Instabilities. Geneva: JACOW; 2016: 2502. DOI: 10.2172/839958

[59] Benedetto E, Schulte D, Zimmermann F, Ohmi K, Papaphilippou Y, Rumolo G. Transverse 'monopole' instability driven by an electron cloud? In: Proceedings of the 2003 Particle Accelerator Conference; 12-16 May 2003; Portland, OR, USA. New York: IEEE; 2003. DOI: 10.1109/PAC.2003.1289811

[60] Win SS, Ohmi K. Coupled-Bunch Instability Caused by Electron Cloud 2004. Geneva: CERN; 2005:313-323. DOI: 10.1109/PAC.2005.1590619

[61] Pivi MTF. CMAD: A new self-consistent parallel code to simulate the electron cloud build-up and instabilities. In: Proceedings of PAC. Albuquerque, New Mexico, USA; 2007. DOI: 10.1109/PAC.2007.4440517

[62] Li K, Cesaratto J, Fox JD, Pivi M, Rivetta C, Rumolo G, editors. Instabilities simulations with wideband feedback systems: CMAD, HEADTAIL, WARP. CERN. 2013;**002**:203-210. DOI: 10.5170/CERN-2013-002.203

[63] Rumolo G, Zimmermann F. Practical user guide for HEADTAIL [Internet]. CERN-SL-Note-2002-036 AP: European organization for nuclear research. 2002. Available from: http://cds.cern.ch/record/702717/files/sl-note-2002-036.pdf [Accessed: 04 February 2002]

[64] Pischer MB, He P, Huang H, Hseuh HC, Iriso U, Rumolo G, Smart L, Trbojevic D, Zhang SY. Electron Clouds and Vacuum Pressure Rise in HIC [Internet]. BNL-72451-2004-CP. 2004. Available from: https://cds.cern.ch/record/846619/files/p53.pdf [Accessed: 04 February 2020]

[65] Valizadeh R, Malyshev OB, Wang S, Zolotovskaya SA, Allan Gillespie W, Abdolvand A. Low secondary electron yield engineered surface for electron cloud mitigation. Applied Physics Letters. 2014;**105**(23):231605. DOI: 10.1063/1.4902993 [Accessed: 04 February 2020]

[66] Costa Pinto P, Calatroni S, Neupert H, Letant-Delrieux D, Edwards P, Chiggiato P, et al. Carbon coatings with low secondary electron yield. Vacuum. 2013;**98**:29-36. DOI: 10.1016/j.vacuum.2013.03.001

[67] Cimino R, Demma T. Electron cloud in accelerators. International Journal of Modern Physics A. 2014;**29**(17):1430023. DOI: 10.1142/S0217751X14300233

[68] Spallino L, Angelucci M, Larciprete R, Cimino R. On the compatibility of porous surfaces with cryogenic vacuum in future high-energy particle accelerators. Applied Physics Letters. 2019;**114**(15):153103. DOI: 10.1063/1.5085754

[69] Arnell RD, Kelly PJ, Bradley JW. Recent developments in pulsed magnetron sputtering. Surface and Coatings Technology. 2004;**188-189**: 158-163. DOI: 10.1016/j.surfcoat.2004.08.010

[70] Le Pimpec F, Kirby RE, King F, Pivi M. Properties of TiN and TiZrV thin film as a remedy against electron cloud. Nuclear Instruments and Methods in Physics Research. 2005;**551**(2-3):187-199. DOI: 10.1016/j.nima.2005.05.048

[71] Todd R, Hseuh HC, Weiss D. Summary on titanium nitride coating of sns ring vacuum chambers. In: Proceedings of 2005 Particle Accelerator Conference, Knoxville, Tennessee. 2005. DOI: 10.1109/PAC.2005.1591373

[72] Le Pimpec F, Kirby RE, King FK, Pivi M. The effect of gas ion bombardment on the secondary electron yield of TiN, TiCN and TiZrV coatings for suppressing collective electron effects in storage rings. Nuclear Instruments and Methods in Physics Research. 2006;**564**(1):44-50. DOI: 10.1016/j.nima.2006.03.041

[73] Variola A, Kaabi W, Jenhani H, Lepercq P, Keppe G, Palmieri V, et al. Titanium nitride coating of rf ceramic windows by reactive dc magnetron sputtering. In: Proceedings of 11th European Particle Accelerator Conference; 23-27 June 2008. Genoa, Italy; 2008. pp. 931-933

[74] Shibata K, Hisamatsu H, Kanazawa K, Suetsugu Y, Shirai M. Development of tin coating

system for beam ducts of kek b-factory. TUPP071. In: Proceedings of EPAC08, 23-27 Jun 2008; Genoa, Italy. Geneva: JACoW; 2008. p. TUPP071

[75] Shibata K, Kanazawa K, Hisamatsu H, Shirai M. Gas desorption from TiN-coated copper beam duct. In: Proceedings of PAC09, 4-8 May 2009. Vancouver, BC, Canada; 2009

[76] Jones MI, Grant DM. Effect of substrate preparation and deposition conditions on the preferred orientation of TiN coatings deposited by RF reactive sputtering. Surface and Coating Technology. 2000;**132**:143-151. DOI: 10.1016/s0257-8972(00)00867-7

[77] Lorkiewicz ABJ, Dwersteg B, Kostin D, Moeller W-D, Layalan M. Surface tin coating of tesla couplers at desy as an antimultipactor remedy. In: The 10th Workshop on RF Superconductivity. Tsukuba, Japan; 2001

[78] Kirby RE. Secondary electron emission yields from PEP-II accelerator materials. Nuclear Instruments and Methods in Physics Research A. 2001;**469**:1-12. DOI: 10.1016/s0168-9002(01)00704-5

[79] Chou WJ, Yu GP, Huang JH. Corrosion behavior of TiN-coated 304 stainless steel. Corrosion Science. 2001;**43**:2023-2035. DOI: 10.1016/s0010-938x(01)00010-5

[80] Chou WJ, Yu GP, Huang JH. Deposition of TiN thin films on Si 100 by HCD ion plating. Surface and Coating Technology. 2001;**140**:206-214. DOI: 10.1016/S0254-0584(03)00209-8

[81] Banerjee R, Chandra R, Ayyub P. Influence of the sputtering gas on the preferred orientation of nanocrystalline titanium nitride thin films. Thin Solid Films. 2002;**405**:64-72. DOI: 10.1016/s0040-6090(01)01705-9

[82] Espinoza-Beltrán FJ, Che-Soberanis O. García-González L, Morales-Hernández J. effect of the substrate bias potential on crystalline grain size, intrinsic stress and hardness of vacuum arc evaporated TiN/c-Si coatings. Thin Solid Films. 2003;**437**(1-2):170-175. DOI: 10.1016/s0040-6090(03)00568-6

[83] Kelly PJ, Beevers CF, Henderson PS, Arnell RD, Bradley JW, Bäcker H. A comparison of the properties of titanium-based films produced by pulsed and continuous DC magnetron sputtering. Surface and Coatings Technology. 2003;**174-175**:795-800. DOI: 10.1016/s0257-8972(03)00356-6

[84] Kennedy K, Harteneck B, Millos G, Benapfl M, King F, Kirby R. TiN coating of the PEP-II low-energy ring aluminum arc vacuum chambers. In: Proceedings of the 1997 Particle Accelerator Conference (Cat. No.97CH36167). 1997. DOI: 10.1109/PAC.1997.753276

[85] Saito K. TiN thin film on stainless steel for extremely high vacuum material. Journal of Vacuum Science & Technology A: Vacuum, Surfaces, and Films. 1995;**13**(3):556. DOI: 10.1116/1.579785

[86] Wang J, Wei W, Zhang B, et al. Simulation of the trajectory of electrons in a magnetron sputtering system of tin with cst particle studio. In: Proceedings of IPAC. Dresden, Germany; 2014. DOI: 10.18429/JACoW-IPAC2014-WEPME036

[87] Wang J, Xu YH, Zhang B, Wei W, Fan L, Pei XT, et al. Experimental study on TiN coated racetrack-type ceramic pipe. Chinese Physics C. 2015;**39**(11):117005

[88] Lorkiewicz J, Brinkmann A, Dwersteg B, Kostin D, Moeller W-D, Layalan M. Surface tin coating of tesla couplers at DESY as an antimultipactor

remedy. In: The 10th Workshop on RF Superconductivity Tsukuba. Japan; 2001

[89] Yamamoto K, Shibata T, Ogiwara N, Kinsho M. Secondary electron emission yields from the J-PARC RCS vacuum components. Vacuum. 2007;**81**(6):788-792. DOI: 10.1016/j.vacuum.2005.11.052

[90] Yin Vallgren C, Bauche J, Calatroni S, et al. Amorphous carbon coatings for mitigation of electron cloud in the CERN SPS. In: Proceedings of 1st International Particle Accelerator Conference, 23-28 May 2010, Kyoto, Japan. 2010. DOI: 10.1103/PhysRevSTAB. 14.071001

[91] Vallgren CY, Ashraf A, Calatroni S, Chiggiato P, Pinto PC, Marques HP, et al. Low secondary electron yield carbon coatings for electron cloud mitigation inmodern particle accelerators. In: Proceedings of 1st International Particle Accelerator Conference; 23-28 May 2010; Kyoto, Japan. Geneva: JACoW; 2010. p. WEOAMH03

[92] Vallgren CY, Chiggiato P, Pinto PC, Neupert H, Rumolo G, Shaposhnikova E, et al. Performance of carbon coatings for mitigation of electron cloud in the SPS. In: Proceedings of IPAC2011; 4-9 Sep 2011, San Sebastián, Spain. 2011. DOI: 10.1007/BF01755788

[93] Cimino R, Angelucci M, Gonzalez LA, Larciprete R. SEY and low-energy SEY of conductive surfaces. Journal of Electron Spectroscopy and Related Phenomena. 2019;**06**:008. DOI: 10.1016/j.elspec.2019.06.008

[94] Prodromides CSAE, Taborelli M. Lowering the activation temperature of TiZrV non-evaporable getter flms. Vacuum. 2001;**60**:35-41. DOI: 10.1016/S0042-207X(00)00243-8

[95] Ferreira MJ, Tallarico DA, Nascente PAP, Paniago RM. Preparation and characterization of Ti-Zr-V non-evaporable getter films to Be used in ultra-high vacuum. In: AIP Conference Proceedings. 2009;**1092**:168-172. DOI: 10.1063/1.3086219

[96] Ferreira MJ, Seraphim RM, Ramirez AJ, Tabacniks MH, Nascente PAP. AIP Conference Proceedings. Characterization and evaluation of Ti-Zr-V non-evaporable getter films used in vacuum systems. Physics Procedia. 2012;**32**:840-852. DOI: 10.1063/1.3086219

[97] Tripathi A, Singh N, Avasthi DK. Hydrogen intake capacity of ZrVFe alloy bulk getters. Vacuum. 1997;**48**(12):1023-1025. DOI: 10.1016/s0042-207x(97)00116-4

[98] Benvenuti C, Cazeneuve JM, Chiggiato P, Cicoira F, Santana AE, Johanek V, et al. A novel route to extreme vacua: The non-evaporable getter thin film coatings. Vacuum. 1999;**53**(1-2):219-225. DOI: 10.1016/s0042-207x(98)00377-7

[99] Lee SM, Park YJ, Lee HY, Kim KC, Baik HK. Hydrogen absorption properties of a Zr-Al alloy ball-milled with Ni powder. Intermetallics. 2000;**8**(7):781-784. DOI: 10.1016/s0966-9795(00)00010-8

[100] Giannantonio R, Succi M, Solcia C. Combination of a cryopump and a non-evaporable getter pump in applications. Journal of Vacuum Science and Technology A. 1997;**15**(1):187-191. DOI: 10.1116/1.580462

[101] Bertolini L, Behne D, Bowman J, Hathaway D, Kishiyama K, Mugge M, et al. Design of the linear non-evaporable getter pump for the PEP-II B factory. In: Proceedings of the 1997 Particle Accelerator Conference. Vol. 1-3. 1998. pp. 3622-3624. DOI: 10.1109/PAC.1997.753294

[102] Benvenuti C, Francia F. Room-temperature pumping characteristics

of a Zr–Al nonevaporable getter for individual gases. Journal of Vacuum Science and Technology A. 1988;**6**:2528. DOI: 10.3969/j.issn.1002-6819.2009.10.062

[103] Kovac J, Sakho O, Manini P, Sancrotti M. Evaluation of temperature-dependent surface-chemistry in Zr2fe and Zrvfe via X-ray photoemission spectroscopy. Surface and Interface Analysis. 1994;**22**(1-12):327-330. DOI: 10.1002/sia.740220171

[104] Chiggiato P. Production of extreme high vacuum with non evaporable getters. Physica Scripta. 1997;**T71**:9-13. DOI: 10.1088/0031-8949/1997/T71/002

[105] Drbohlav J, Matolin V. Static SIMS study of Ti, Zr, V and Ti-Zr-V NEG activation. Vacuum. 2003;**71** (1-2):323-327. DOI: 10.1016/s0042-207x(02)00757-1

[106] Drbohlav J, Matolinova I, Masek K, Matolin V. Sims study of Ti-Zr-VNEG thermal activation process. Vacuum. 2005;**80**(1-3):47-52. DOI: 10.1016/j.vacuum.2005.07.016

[107] Li CC, Huang JL, Lin RJ, Chen CH, Lii DF. Characterization of activated non-evaporable porous Ti and Ti-Zr-V getter films by synchrotron radiation photoemission spectroscopy. Thin Solid Films. 2006;**515**(3): 1121-1125. DOI: 10.1016/j.tsf.2006.07.052

[108] Sutara F, Skala T, Masek K, Matolin V. Surface characterization of activated Ti-Zr-V NEG coatings. Vacuum. 2009;**83**(5):824-827. DOI: 10.1016/j.vacuum.2008.08.002

[109] Prodromides AE, Taborelli M. The characterisation of non-evaporable getters by auger electron spectroscopy: Analytical potential and artefacts. Applied Surface Science. 2002;**191**:300-312. DOI: 10.1016/s0169-4332(02)00222-2

[110] Malyshev OB, Gurran L, Goudket P, Marinov K, Wilde S, Valizadeh R, et al. RF surface resistance study of non-evaporable getter coatings. Nuclear Instruments and Methods in Physics Research. 2017;**844**:99-107. DOI: 10.1016/j.nima.2016.11.039

[111] Malyshev OB, Valizadeh R, Hannah AN. Pumping and electron-stimulated desorption properties of a dual-layer nonevaporable getter. Journal of Vacuum Science & Technology A: Vacuum, Surfaces, and Films. 2016;**34**(6):061602. DOI: 10.1116/1.4964612

[112] Malyshev OB, Valizadeh R, Hogan BT, Hannah AN. Electron-stimulated desorption from polished and vacuum fired 316LN stainless steel coated with Ti-Zr-Hf-V. Journal of Vacuum Science & Technology A: Vacuum, Surfaces, and Films. 2014;**32**(6):061601. DOI: 10.1116/1.4897932

[113] Calder RS. Ion induced gas desorption problems in the ISR. Vacuum. 1974;**24**:437-443. DOI: 10.1016/0042-207x(74)90001-3

[114] Marie-Hélène A, Calder R, Mathnwson A. The effect of bakeout temperature on the electron and ion induced gas desorption coefficients of some technological materials. Vacuum. 1978;**29**(2):53-65. DOI: 10.1016/S0042-207X(79)80335-8

[115] Baraglola RA. Ion-induced desorption of surface contaminants. Journal of Nuclear Materials. 1984; **126**:313-316. DOI: 10.1016/0022-3115(84)90043-6

[116] Rossi A, Malyshev OB. Ion desorption stability in the LHC. Vacuum Technical Note 99-20; 1999

[117] Malyshev OB. The ion impact energy on the LHC vacuum chamber walls. In: Proceedings of the 7th European Particle

Accelerator Conference; 26-30 Jun 2000. Vienna, Austria; 2000

[118] Malyshev OB, Rossi A. Ion desorption vacuum stability in the LHC the multigas model. In: Proceedings of the 7th EPAC; 26-30 Jun 2000. Vienna, Austria; 2000

[119] Lozano MP. Ion-induced desorption yield measurements from copper and aluminium. Vacuum. 2002;**67**:339-345. DOI: 10.1016/s0042-207x(02)00223-3

[120] Li P, Yang JC, Dong ZQ, Zheng WH, Xie WJ, Chang JJ, et al. Beam loss simulation and gas desorption measurement for HIAF. In: Proceedings of the 13th Symposium on Accelerator Physics. HuNan, China; 2017

[121] Malyshev OB. Ion induced pressure instability in the ILC positron DR. In: Proceedings of IPAC'10; 23-28 May 2010. Kyoto, Japan; 2010

[122] Maurer C, Hoffmann DHH, Bozyk LHJ, Kollmus H, Spiller PJ. Heavy ion induced desorption measurements on cryogenic targets. In: Proceedings of the 5th International Particle Accelerator Conference; 16-20 June 2014. Dresden, Germany; 2014

[123] Christoph M, Holger K, et al. Simulation and experimental investigation of heavy ion induced desorption from cryogenic targets. In: Proceedings of IPAC; 2015; Richmond, VA. USA: THPF; 2015. pp. 3699-3701. DOI: 10.18429/JACoW-IPAC2015-THPF010

[124] Malyshev OB, Collins IR. Estimates of Photon Induced Gas Densities in the Long Straight Sections of IR1 and IR5. 1999. Geneva: CERN-LHC-VAC. 1999;**99**:14. DOI: 10.4028/www.scientific.net/AMM.278-280.831

[125] Anashin VV, Collins IR, Dostovalov RV, Korotaeva ZA, Krasnov AA, Malyshev OB, et al. Vacuum performance of a carbon fibre cryosorber for the LHC LSS beam screen. Vacuum. 2004;**75**(4):293-299. DOI: 10.1016/j.vacuum.2004.03.010

[126] Anashin VV, Dostovalov RV, Krasnov AA, Collins IR, Malyshev OB. Vacuum performance of a beam screen with charcoal for the LHC long straight sections. Vacuum. 2004;**72**(4):379-383. DOI: 10.1016/j.vacuum.2003.09.006

Chapter 4

Very Compact Linear Colliders Comprising Seamless Multistage Laser-Plasma Accelerators

Kazuhisa Nakajima, Min Chen and Zhengming Sheng

Abstract

A multistage laser-plasma accelerator (LPA) driven by two mixing electromagnetic hybrid modes of a gas-filled capillary waveguide is presented. Plasma wakefields generated by a laser pulse comprising two mixing modes coupled to a metallic or dielectric capillary filled with gas provide us with an efficient accelerating structure of electrons in a substantially long distance beyond a dephasing length under the matching between a capillary radius and plasma density. For a seamless multistage structure of the capillary waveguide, the numerical model of the transverse and longitudinal beam dynamics of an electron bunch considering the radiation reaction and multiple Coulomb scattering effects reveals a converging behavior of the bunch radius and normalized emittance down to $\sim$1 nm level when the beam is accelerated up to 560 GeV in a 67 m length. This capability allows us to conceive a compact electron-positron linear collider providing with high luminosity of 10^{34} cm^{-2} s^{-1} at 1 TeV center-of-mass (CM) energy.

Keywords: future colliders, lepton colliders, laser-plasma accelerators, multistage coupling, CAN lasers

1. Introduction

In the long-standing quest for the fundamental building blocks of nature, the so-called Standard Model of particle physics, energy frontier colliders have played a central role in the forefront research for matter and interactions. For future high-energy particle colliders to explore physics beyond the Standard Model, a proton-proton circular collider at energy of 100 TeV in a 100 km circumference or electron-positron linear collider with energy of the order of 1 TeV in a 30 km length is being considered around the world, exploiting the conventional technologies such as superconducting magnets or RF systems [1]. In contrast to proton colliders that create clouds of debris, electron-positron colliders enable cleaner and more precision experiments of fundamental particle collisions. Nowadays, a diversity of electron-positron linear colliders is proposed as a potential application of advanced accelerator concepts [2], such as two beam accelerators, dielectric wakefield accelerators, beam-driven plasma wakefield accelerators, and laser-driven plasma wakefield accelerators [3], promising with much higher accelerating gradients than that of a conventional RF accelerator.

Laser-plasma accelerators (LPAs) [4, 5] can support a wide range of potential applications requiring high-energy and high-quality electron-positron beams. In particular, field gradients, energy conversion efficiency, and repetition rates are essential factors for practical applications such as compact free electron lasers [6, 7] and high-energy frontier colliders [8, 9]. Although LPAs provide enormous accelerating gradients, as high as 100 GV/m at the plasma density of 10^{18} cm^{-3}, dephasing of relativistic electrons with respect to a correct acceleration phase of the plasma wakefield with the phase velocity that is smaller than the speed of light in vacuum, and energy depletion of the laser pulse limit the electron energy gain in a single stage. A straightforward solution to overcome the dephasing and pump depletion effects is to build a multistage accelerator comprising consecutive LPA stages [3] such that a final energy gain reaches the requirement of the beam energy without loss of the beam charge and qualities through a coupling segment where a fresh laser pulse is fed to continuously accelerate the particle beam from the previous stage. The propagation of laser pulses in plasmas is described by refractive guiding, in which the refractive index can be modified from the linear free space value mainly by relativistic self-focusing, ponderomotive channeling, and a preformed plasma channel [10]. The self-guided LPA [11–14] relies only on intrinsic effects of relativistic laser-plasma interactions such as relativistic self-focusing and ponderomotive channeling. On the other hand, the channel-guided LPA exploits a plasma waveguide with a preformed density channel [15–17] or a gas-filled capillary waveguide made of metallic or dielectric materials [18]. The plasma waveguide is likely to propagate a single-mode laser pulse through a radially parabolic distribution of the refractive index and generates plasma waves inside the density channel, the properties of which are largely affected by a plasma density profile and laser power [19]. In contrast with plasma waveguides, the capillary waveguide can guide the laser due to Fresnel reflection on the inner capillary wall, and plasma waves are generated in an initially homogeneous plasma, relying on neither laser power nor plasma density. The presence of the modal structure imposed by the boundary conditions at the capillary wall affects the propagation of a laser pulse through the capillary and thus the excitation of plasma waves inside the capillary. This characteristic allows us to control acceleration of electrons through the modal structure of the propagation of the laser pulse as long as the laser intensity on the capillary wall is kept below the material breakdown [20, 21].

In this paper, we present a novel scheme of a gas-filled capillary accelerator driven by a laser pulse formed from two-mode mixing of the capillary eigenmodes, so-called electromagnetic hybrid modes [20]. Two coupled eigenmodes with a close longitudinal wave number can generate beating wakefields in the capillary. When the beating period is equal to the dephasing distance, the electrons experience the rectified accelerating field; thereby their energy gain can increase over many accelerating phases exceeding the linear dephasing limit and reach the saturation due to the energy depletion of a drive laser pulse in the single-stage LPA. For efficient acceleration of the electron-positron beam up to an extremely high energy such as TeV energies, the multistage accelerator comprising a series of plasma-filled capillary waveguides is a sound approach, in which the particle beam is injected into the initial stage at the right phase of the wakefield from the external injector and accelerated cumulatively in the consecutive accelerating phase of successive stages. For applications of extreme high-energy particle beams to TeV center-of-mass (CM) energy electron-positron linear colliders, minimizing the transverse normalized emittance of the beam particles is of essential importance to meet the requirement of the luminosity of the order of 10^{34} cm^{-2} s^{-1} at 1 TeV CM energy for the particle physics experiments [22]. The numerical model on the bunched beam dynamics in laser wakefields, based on the exact solution of single particle betatron motion taking into account the radiation reaction and multiple Coulomb scattering, reveals that the transverse normalized emittance and beam radius can be consecutively reduced during continuous acceleration in the presence of optimally phased recurrence of longitudinal and transverse wakefields [19]. The final properties of the particle beams reached to the objective energy meet the requirements of the luminosity without resort to an additional focusing system.

The remaining part of this paper is organized as follows. In Section 2, the complete description on the longitudinal and transverse laser wakefields generated by two electromagnetic hybrid modes with moderate intensities coupled to a gas-filled capillary waveguide is provided. In Section 3, the particle beam dynamics on energy gain, beam loading, and betatron motion in a single stage of the two-mode mixing LPA is investigated, taking into account radiation reaction and multiple Coulomb scattering with plasma ions. In Section 4, a multistage coupling with a variable curvature plasma channel is presented. For the multistage comprising two-mode mixing LPAs, the results of numerical studies on the transverse beam dynamics of a particle bunch are shown. Analytical consideration on the evolution of the normalized emittance of the particle beam in the presence of radiation reaction and the multiple Coulomb scattering is given. In Section 5, the performance of a 1-TeV CM energy electron-positron collider comprising the multistage two-mode mixing LPAs is discussed on the luminosity and beam-beam interaction. In Section 6, we conclude our investigation on the proposed laser-plasma linear collider with a summary.

2. Laser pulse propagation in a gas-filled capillary tube

For a large-scale accelerator complex such as the energy frontier particle beam colliders, it is axiomatically useful in assembling a long-range multistage structure for the use of long-term experimental operation at a high-precision and high-repetition rate that each electromagnetic waveguide consists of a simple monolithic structure, as referred to the design of the future electron-positron linear colliders based on radio-frequency technologies [22]. Despite the long-standing research on plasma waveguides comprising density channels generated in plasmas with laser-induced hydrodynamic expansion [23, 24] and pulsed discharges of an ablative capillary [25, 26] or a gas-filled capillary [27, 28], a length of such a plasma channel has been limited to about 10 cm. The pulsed discharge capillaries relying on collisional plasma processes have some difficulties in plasma densities less than 10^{17} cm^{-3} and the temporal and spatial stabilities of the density channel properties for the operation at a high repletion rate such as 10 kHz [5, 29]. In contrast to pulsed discharge plasma waveguides, metallic or dielectric capillary waveguides filled with gas [18, 30] will be revisited for a large-scale laser-plasma accelerator operated at a practically higher-repetition rate than 10 kHz, because of the passive optical guiding of laser pulses, the propagating electromagnetic fields of which are simply determined the boundary conditions on a static solid wall of the waveguide unless the laser intensity is high enough to cause the material breakdown on a capillary wall [20, 21]. Furthermore, the modal nature of electromagnetic fields arising from the boundary conditions on a solid wall allows us to conceive a novel scheme that can overcome a drawback of LPAs, referred to as dephasing of accelerated electron beams from a correct acceleration phase in laser wakefields.

2.1 Laser-driven wakefields generated by two capillary modes

Considering the electromagnetic hybrid modes EH_{1n} [20] to which the most efficient coupling of a linearly polarized laser pulse in vacuum occurs, the normalized vector potential for the eigenmode of the n-th order is written by [31].

$$a_n = a_{n0}J_0(u_n r/R_c)\exp\left[-k_n^l z - \frac{(z-v_{g,n}t)^2}{2c^2\tau^2}\right]\cos(\omega_0 t - k_{zn}z), \tag{1}$$

where a_{n0} is the amplitude of the normalized vector potential defined as $a_{n0} \equiv eA_{n0}/m_e c^2$ for the EH_{1n} mode with the vector potential A_{n0}, the electron charge e, electron mass m_e, and the speed of light in vacuum c; J_0 the zero-order Bessel function of the first kind; u_n the n-th zero of J_0; r the radial coordinate of the capillary in cylindrical symmetry; R_c the capillary radius; z the longitudinal coordinate; τ the pulse duration; and ω_0 the laser frequency. The longitudinal wave number k_{zn}, the damping coefficient k_n^l, and the group velocity of the n-th mode $v_{g,n}$ are given by [20].

$$k_{zn} = \left(k_0^2 - \frac{u_n^2}{R_c^2}\right)^{1/2}, k_n^l = \frac{u_n^2(1+\varepsilon_r)}{2k_{zn}^2 R_c^3(\varepsilon_r - 1)^{1/2}}, v_{g,n} \simeq c\left(1 - \frac{u_n{}^2}{k_0^2 R_c{}^2}\right)^{1/2}, \tag{2}$$

where $k_0 = \omega_0/c = 2\pi/\lambda_0$ is the laser wavenumber with the laser wavelength λ_0 and ε_r is the relative dielectric constant. In the quasi-linear wakefield regime $|\mathbf{a}| = e|\mathbf{A}|/(m_e c^2) \sim 1$, the ponderomotive force exerted on plasma electrons by two coupled capillary laser fields $\mathbf{a}_{nm} = \mathbf{a}_n + \mathbf{a}_m$ can be written by $\mathbf{F}_p = -m_e c^2 \beta_g \nabla a_{nm}^2/2$, where a_{nm}^2 is defined by averaging the nonlinear force over the laser period $2\pi/\omega_0$, i.e., assuming that $v_{g,n} \sim v_{g,m} \sim v_g$ in the propagation distance $z \le z_{\text{mix}} \approx 8\pi^2(R_c/\lambda_0)^2 c\tau/(u_m{}^2 - u_n{}^2)$, where z_{mix} is the mode mixing length over which two hybrid modes EH_{1n} and EH_{1m} overlap to cause the beatings of the normalized vector potential, e.g., $z_{\text{mix}} \sim 56$ cm for the EH_{11} - EH_{12} mode mixing of a laser pulse with $\tau = 25$ fs and $\lambda_0 = 1\ \mu$m in a capillary tube with $R_c = 152.6\ \mu$m

$$\begin{aligned} a_{nm}{}^2(r,t) &= \frac{1}{2}a_{n0}{}^2 J_0{}^2\left(\frac{u_n r}{R_c}\right)\exp\left[-2k_n^l z - \frac{(z-v_g t)^2}{c^2\tau^2}\right] + \frac{1}{2}a_{m0}{}^2 J_0{}^2\left(\frac{u_m r}{R_c}\right)\exp\left[-2k_m^l z - \frac{(z-v_g t)^2}{c^2\tau^2}\right] \\ &+ a_{n0}a_{m0}J_0\left(\frac{u_n r}{R_c}\right)J_0\left(\frac{u_m r}{R_c}\right)\exp\left[-\left(k_n^l + k_m^l\right)z - \frac{(z-v_g t)^2}{c^2\tau^2}\right]\cos(k_{zm} - k_{zn})z. \end{aligned} \tag{3}$$

The electrostatic potential $\Phi(r,t)$ defined by $\mathbf{F}(r,t) = -e\nabla\Phi(r,t)$ is obtained from Eq. (5).

$$\left(\frac{\partial^2}{\partial t^2} + \omega_p^2\right)\Phi(r,t) = \frac{\omega_p^2 m_e c^2 \beta_g}{2e} a_{nm}^2(r,t) \tag{4}$$

where $\omega_p = (4\pi e^2 n_e/m_e)^{1/2}$ is the plasma frequency. The solution of Eq. (4) is

$$\begin{aligned} \Phi(r,t) &= \frac{\sqrt{\pi}}{8}\left(\frac{m_e c^2}{e}\right)\beta_g k_p c\tau e^{-\left(\frac{k_p c\tau}{2}\right)^2}\left[a_{n0}{}^2 J_0{}^2\left(\frac{u_n r}{R_c}\right)e^{-2k_n^l z} + a_{m0}{}^2 J_0{}^2\left(\frac{u_m r}{R_C}\right)e^{-2k_m^l z}\right. \\ &\left.+ 2a_{n0}a_{m0}J_0\left(\frac{u_n r}{R_c}\right)J_0\left(\frac{u_m r}{R_c}\right)e^{-(k_n^l + k_m^l)z}\cos\Delta k_{znm}z\right]\left[S(z)\cos k_p(z - v_g t) + C(z)\sin k_p(z - v_g t)\right], \end{aligned} \tag{5}$$

where $k_p = \omega_p/v_g$ is the plasma wavenumber in the capillary, $\beta_g = v_g/c$, $\Delta k_{znm} = k_{zn} - k_{zm}$ the mode beating wavenumber and

$$C(z) = \Re\mathrm{erf}\left(\frac{z - v_g t}{c\tau} + i\frac{k_p c\tau}{2}\right) - 1, \; S(z) = \Im\mathrm{erf}\left(\frac{z - v_g t}{c\tau} + i\frac{k_p c\tau}{2}\right), \tag{6}$$

with the real ($\Re$) and imaginary ($\Im$) part of the error function $\mathrm{erf}(z) = (2/\sqrt{\pi})\int_0^z e^{-s^2} ds$ [5]. For k_n^l, $k_m^l \ll k_p$ and $\Delta k_{znm} \ll k_p$, the longitudinal electric field generated by the laser pulse can be obtained from $E_{zL} = -\partial\Phi/\partial z$ as

$$E_{zL}(r,z,t) = \frac{\sqrt{\pi}}{8}\left(\frac{m_e c\omega_p}{e}\right) k_p c\tau e^{-\left(\frac{k_p c\tau}{2}\right)^2}\left[a_{n0}^2 J_0\left(\frac{u_n r}{R_c}\right)e^{-2k_n^l z} + a_{m0}^2 J_0\left(\frac{u_m r}{R_c}\right)e^{-2k_m^l z}\right.$$
$$\left. +2a_{n0}a_{m0}J_0\left(\frac{u_n r}{R_c}\right)J_0\left(\frac{u_m r}{R_c}\right)e^{-\left(k_n^l + k_m^l\right)z}\cos\Delta k_{znm}z\right]\left[S(z)\sin k_p(z - v_g t) - C(z)\cos k_p(z - v_g t)\right]. \tag{7}$$

The transverse focusing force is obtained from $F_{rL} = e(E_r - B_\varphi) = -\partial\Phi/\partial r$ as

$$\begin{aligned} F_{rL}(r,z,t) = {} & \frac{\sqrt{\pi}}{4}\left(\frac{m_e c\omega_p}{e}\right)\frac{c\tau}{R_c} e^{-\left(\frac{k_p c\tau}{2}\right)^2}\left[a_{n0}{}^2 u_n J_0\left(\frac{u_n r}{R_c}\right)J_1\left(\frac{u_n r}{R_c}\right)e^{-2k_n^l z} + a_{m0}{}^2 u_m J_0\left(\frac{u_m r}{R_c}\right)J_1\left(\frac{u_m r}{R_c}\right)e^{-2k_m^l z}\right. \\ & \left. + a_{n0}a_{m0}\left(u_n J_0\left(\frac{u_m r}{R_c}\right)J_1\left(\frac{u_n r}{R_c}\right) + u_m J_0\left(\frac{u_n r}{R_c}\right)J_1\left(\frac{u_m r}{R_c}\right)\right)e^{-\left(k_n^l + k_m^l\right)z}\cos\Delta k_{znm}z\right] \\ & \times\left[C(z)\sin k_p(z - v_g t) + S(z)\cos k_p(z - v_g t)\right], \end{aligned} \tag{8}$$

where $J_1(z) = -J_0'(z)$ is the Bessel function of the first order.

The proposed scheme restricts the laser intensity such that the plasma response is within the quasi-linear regime, i.e., $a_0 \sim 1$, for two reasons. The one is avoidance of the nonlinear plasma response such as in the bubble regime, where symmetric wakefields for the electron and positron beams cannot be obtained for the application to electron-positron colliders [8, 9] and the degradation of the beam quality due to the self-injection of dark currents from the background plasma electrons. The other is an inherent demand that the laser intensity guided in a capillary tube should be lower enough than the threshold of material damage on the capillary wall [19].

2.2 Coupling control for generating two capillary modes

The coupling efficiency C_n defined by an input laser energy with a spot radius r_0 and amplitude a_0 coupled to the E_{1n} mode in the capillary with the radius R_c, i.e., $a_{n0}^2 = C_n a_0^2$ is calculated for a linearly polarized Airy beam,

$$C_n = \frac{4}{J_1^2(u_n)}\left[\int_0^1 J_1\left(\frac{\nu_1 R_c x}{r_0}\right)J_0(u_n x)dx\right]^2, \tag{9}$$

and for a Gaussian beam,

$$C_n = 8\left(\frac{R_c}{r_0 J_1(u_n)}\right)^2\left[\int_0^1 x\exp\left(-\frac{x^2 R_c^2}{r_0^2}\right)J_0(u_n x)dx\right]^2, \tag{10}$$

where $\nu_1 = 3.8317$ is the first root of the equation of $J_1(x) = 0$ [20], as shown in **Figure 1a** and **b**, respectively, as a function of R_c/r_0. In Eq. (5), the beating term can be maximized by setting R_c/r_0 at which $(C_n C_m)^{1/2}$ has the maximum value and the minimum fraction of higher-order modes. As shown in **Figure 1**, the Airy beam generates the maximum EH_{11}-EH_{12} mode mixing with $(C_1 C_2)^{1/2} = 0.45$ and a fraction of higher-order modes with ~0.5% at $R_c/r_0 = 1.67$, where the coupling

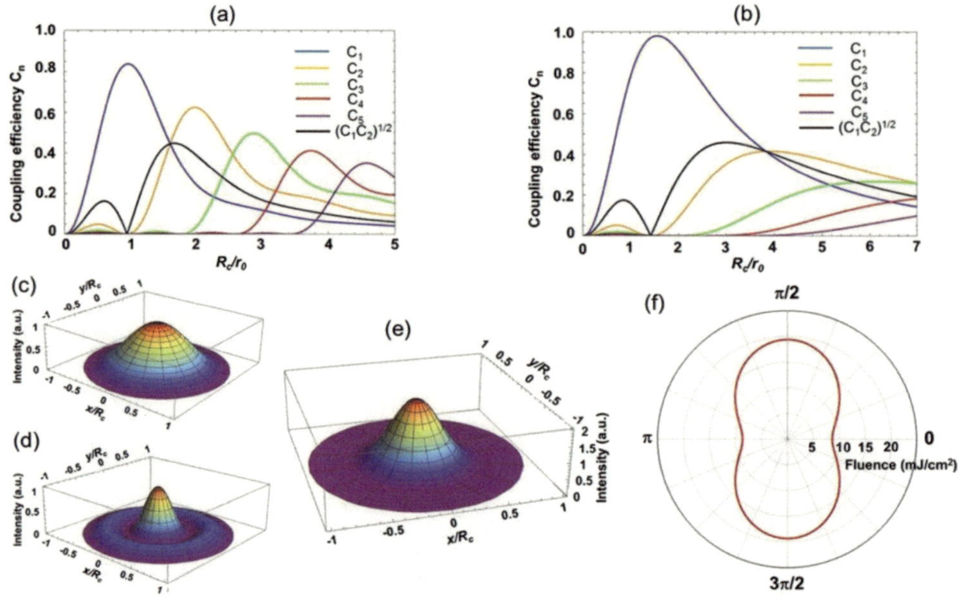

Figure 1.
(a and b) coupling efficiency C_n for an airy beam and a Gaussian beam with a spot radius r_o, respectively, coupled to the electromagnetic hybrid mode EH_{1n} in a capillary tube with a radius R_c. (c, d, and e) radial intensity profiles for the EH_{11}, EH_{12} monomode, and EH_{11}-EH_{12} mixing mode for the airy beam case. (f) Energy fluence traversing the capillary wall on $R_c = 152.6\ \mu m$ for the peak intensity $I_L = 1.37 \times 10^{18}\ W/cm^2$ ($a_o^2 = 1$) and the pulse duration $\tau_L = 25\ fs$.

efficiencies are $C_1 = 0.4022$, $C_2 = 0.4986$, $C_3 = 0.002366$, $C_4 = 0.001219$, and $C_5 = 0.000701$. The Gaussian beam can generate the EH_{11}-EH_{12} mode mixing with $(C_1C_2)^{1/2} = 0.46$ and a fraction of higher-order modes with ~5.1% at $R_c/r_0 = 3.0$, where the coupling efficiencies are in the order of $C_1 = 0.5980$, $C_2 = 0.3531$, $C_3 = 0.04706$, $C_4 = 0.001815$, and $C_5 = 0.000022$.

The radial intensity profiles for the EH_{11}, EH_{12} monomode and EH_{11}-EH_{12} mixing mode for the Airy beam case are illustrated in **Figure 1c–e**, respectively. As shown in **Figure 1e**, a high-intensity region of the mixing mode is confined within a half radius of the capillary, compared to the monomode intensity profiles, which have a widespread robe toward the capillary wall. A centrally concentrated intensity profile of the mixing mode considerably decreases the energy flux traversing on the capillary wall. The normalized flux for EH_{1n} mode at the capillary wall depends on the azimuthal angle θ as $F_w^n = [u_nJ_1(u_n)/k_0R_c]^2(\cos^2\theta + \varepsilon_r \sin^2\theta)/(\varepsilon_r - 1)^{1/2}$, defined by the ratio of the radial component of the Poynting vector at $r = R_c$ to the longitudinal component of the on-axis Poynting vector [20]. For the Airy beam with $\lambda_0 = 1\ \mu$m coupled to the capillary with $\varepsilon_r = 2.25$ and $R_c = 152.6\ \mu$m, the maximum normalized fluxes for the EH_{11}, EH_{12} mono- and EH_{11}-EH_{12} mixing modes at $\theta = \pi/2$ or $3\pi/2$ are 1.37×10^{-6}, 3.85×10^{-6}, and 6.26×10^{-7}, respectively. The energy fluence traversing the capillary wall can be estimated by $\mathcal{F}_{\text{wall}} \sim F_w^n I_L \tau_L$ for the peak intensity $I_L = 1.37 \times 10^{18}\ \text{W/cm}^2$ ($a_0^2 = 1$) and the pulse duration $\tau_L = 25$ fs, providing the maximum fluences 19, 66, and 19 mJ/cm^2 for the corresponding modes, as shown in **Figure 1f**. The experimental study of laser-induced breakdown in fused silica (SiO_2) [32] suggests that the fluence breakdown threshold is scaled to be $\mathcal{F}_{\text{th}} \sim 120 - 160\ \text{J/cm}^2$ for $\tau_L = 25$ fs. According to a more detailed study of laser propagation in dielectric capillaries under non-ideal coupling conditions [33], the threshold intensity for wall ionization is obtained as $I_{\text{th}} \sim 2.3 \times 10^{18}\,\text{W/cm}^2$ ($a_0 \simeq 1.3$) at the wavelength $\lambda_0 = 1\mu$m for the capillary radius $R_c = 152.6\ \mu$m.

The coupling efficiency of an incident laser pulse to a capillary tube filled with plasma can be improved by the use of a cone-shape entrance of the capillary [34], suppressing self-focusing effects and increasing the accelerating wakefield excited in the capillary. For the propagation of a laser beam with an approximately Gaussian intensity profile $|a|^2 \sim a_0^2 (r_0^2/r_s^2) \exp(-2r^2/r_s^2)$, the evolution of a normalized spot radius $R = r_s/r_0$ can be obtained from the equation $d^2R/dz^2 = 1/(Z_R^2 R^3)(1 - P/P_c)$ [35], where $Z_R = k_0 r_0^2/2$ is the vacuum Rayleigh length, P the laser power, and P_c the critical power for relativistic self-focusing with $P/P_c = k_p^2 r_0^2 a_0^2/32$. For the coupling of an Airy beam (or a Gaussian beam) with the radius $r_0 = R_c/1.67$ ($R_c/3$) to the capillary tube filled with plasma at the electron density of $n_e = 1 \times 10^{18}\ \text{cm}^{-3}$, the cone with the opening radius of $r_i = r_0 (P/P_c)^{1/2} \sim 3r_0$ ($1.7r_0$) and length $z_c = (Z_R/2)(P/P_c - 1)^{1/2} \sim 1.43 Z_R$ ($\sim 0.68 Z_R$) can effectively guide and collect the incident laser energy. The effect of the relativistic self-focusing is estimated by considering the modulation of the refractive index for the EH_{1n} mode, i.e., $\eta_n = 1 - \omega_p^2/(2\omega_0^2)(1 - \delta\phi + \Delta_n) - u_n^2/(2k_0^2 R_c^2)$ [31], where $\delta\phi = e\Phi/m_e c^2 \simeq a^2/4 - \delta n/n_0$ [36] and $\Delta_n \sim 3C_n^2/32 - 5C_n^3/128$. The maximum modulation due to the relativistic self-focusing effect is at most 0.5% for the propagation of the EH_{11}-EH_{12} mixing modes in a capillary.

3. Beam dynamics in a single-stage two-mode mixing LPA

3.1 Electron acceleration

In the linear wakefields excited by two coupled modes EH_{11} and EH_{12} in the capillary waveguide, the longitudinal motion of an electron traveling along the capillary axis at a normalized velocity $\beta_z = v_z/c \approx 1$ is described as [5].

$$d\gamma/dz = -k_p E_{z0}/E_0,\ d\Psi/dz \approx k_p \left(1 - \beta_g\right) \approx k_p/\left(2\gamma_g^2\right), \tag{11}$$

where $m_e c^2 \gamma$ is the electron energy, $E_{z0} = E_{zL}(0, z, t)$ the accelerating field at $r = 0$, $E_0 = m_e c \omega_p/e$ the nonrelativistic wave-breaking field, $\Psi \simeq k_p \left(z - v_g \int_0^z dz/v_z\right)$ the particle phase with respect to the plasma wave, and $\gamma_g = \left(1 - \beta_g^2\right)^{-1/2} \gg 1$. Here, the phase-matching condition is determined such that the beating wavelength is equal to the dephasing length, i.e., $L_{dp} = \lambda_p \gamma_g^2/2 = \pi/(2\Delta k_{z12})$

$$\Delta k_{z12} = k_{z1} - k_{z2} \approx \left(u_2^2 - u_1^2\right)/\left(2k_0 R_c^2\right) = k_p/\left(2\gamma_g^2\right). \tag{12}$$

Taking into account $C(z) \to -2$ and $S(z) \to 0$ for $z - v_g t \ll -c\tau$ and setting the pulse duration of a drive laser pulse with a Gaussian temporal profile to be the optimum length $k_p c\tau = \sqrt{2}$, the on-axis accelerating field near the matching condition is given by

$$E_{z0}/E_0 = -\sqrt{\pi/8}\,a_0{}^2 e^{-\left(1+4\alpha_d a_0^2 \Psi\right)/2} \left[C_1 + C_2 + 2\sqrt{C_1 C_2} \cos(\Psi + \delta)\right] \cos\Psi, \tag{13}$$

where $\delta = \left(\Delta k_{z12} - k_p/2\gamma_g^2\right)L_{dp} \sim \pi\left[\left(u_2^2 - u_1^2\right)\gamma_g^3/\left({k_0}^2R_c^2\right) - 1\right]/2$ is a phase mismatching.

While propagating through plasma and generating wakefields, the laser pulse loses its energy as $\partial\mathcal{E}_L/\partial z \sim -\mathcal{E}_L/L_{\rm pd}$ [37] where $L_{\rm pd}$ is the characteristic scale length of laser energy deposition into plasma wave excitation, referred to as the pump depletion length. In the linear wakefield regime where a laser pulse duration is assumed to be fixed, the laser energy evolution in the capillary can be written as $\mathcal{E}_L(z) \propto a_0^2(C_1 + C_2)e^{-z/L_{\rm pd}-2\left(k_1^l+k_2^l\right)z}$, taking into account the energy attenuation of two coupled hybrid modes. In the quasi-linear wakefield regime, i.e., $a_0^2 \leq 1$, the scaled pump depletion length is given by $k_pL_{\rm pd} = \gamma_g^2/\left(\alpha_d a_0^2\right)$ with $\alpha_d \simeq (C_1 + C_2)/17.4$ for a Gaussian laser pulse [9, 37], while the scaled coupled mode attenuation length yields $k_p/2\left(k_1^l + k_2^l\right) \sim 0.35{\gamma_g}^{7/2}u_2 \gg k_pL_{\rm pd}$ with the matching condition given by Eq. (12), i.e., $k_pR_c = \gamma_g^{1/2}\left(u_2^2 - u_1^2\right)^{1/2}$ for $u_2 > u_1$ and the glass with the relative dielectric constant $\varepsilon_r = 2.25$. Hence, the damping of wakefields during the laser pulse propagation is dominated by the energy depletion of the laser pulse as given in Eq. (13). Thus, integrating the equations of motion in Eq. (11) over $\Psi_0 \leq \Psi \leq \Psi_{\rm mix}$, the energy $m_ec^2\gamma$ and phase of the electron can be obtained as

$$\gamma(\Psi) = \gamma_0 + G(\Psi) - G(\Psi_0), \Psi(z) = \Psi_0 + k_pz/\left(2\gamma_g^2\right), \tag{14}$$

where $m_ec^2\gamma_0$ is the initial electron energy, Ψ_0 the initial electron phase with respect to the wakefield, $\Psi_{\rm mix} = \Psi(z_{\rm mix}) = \Psi_0 + \sqrt{2}\gamma_g$ the maximum electron phase in the wakefield for the matching condition in Eq. (12) and the laser pulse length $k_pc\tau = \sqrt{2}$, and

$$\begin{aligned} G(\Psi) = \gamma_g^2\sqrt{\frac{\pi}{2}}{a_0}^2(C_1 + C_2)e^{-\left(1+4\alpha_d a_0^2\Psi\right)/2}\Bigg[&\frac{\sin\Psi - 2\alpha_d a_0^2\cos\Psi}{1 + 4\alpha_d^2a_0^4} \\ &- \frac{\sqrt{C_1C_2}}{2(C_1 + C_2)}\left(\frac{\cos\delta}{\alpha_d a_0^2} - \frac{\sin(2\Psi + \delta) - \alpha_d a_0^2\cos(2\Psi + \delta)}{1 + \alpha_d^2a_0^4}\right)\Bigg]. \end{aligned} \tag{15}$$

The maximum energy gain to be attainable at $\Psi \to \infty$ is obtained as

$$\begin{aligned} \Delta\gamma_{\rm max}(\delta) &= \gamma_{\rm max} - \gamma_0 = G(\infty) - G(0) \\ &= \gamma_g^2\sqrt{\frac{\pi}{2}}{a_0}^2(C_1 + C_2)e^{-1/2}\left[\frac{2\alpha_d a_0^2}{1 + 4\alpha_d^2a_0^4} + \frac{\sqrt{C_1C_2}}{2(C_1 + C_2)}\left(\frac{\cos\delta}{\alpha_d a_0^2} - \frac{\sin\delta - \alpha_d a_0^2\cos\delta}{1 + \alpha_d^2a_0^4}\right)\right]. \end{aligned} \tag{16}$$

Considering the mixing of two lowest order hybrid modes EH_{11} and EH_{12} with the coupling efficiencies $C_1 = 0.4022$ and $C_2 = 0.4986$, the evolution of the energy gain with respect to γ_g^2 is shown in **Figure 2a** for various detuning phases δ in comparison with that of the EH_{11} monomode with $C_1 = 1$ and $C_2 = 0$. The effect of phase mismatching on the maximum attainable energy gain is shown in **Figure 2b** for various normalized laser intensities a_0^2 in the quasi-linear regime. One can see that the growth of energy gain occurs in the relatively wide range of the phase mismatching over $-\pi/2 \leq \delta \leq \pi/2$ and that the maximum attainable energy gain does not strongly depend on the normalized vector potential a_0 in the quasi-linear regime. While the single-mode LPA driven by the normalized intensity $a_0^2 = 1$ reaches the maximum energy gain $\Delta\gamma_{\rm max} = 0.71\gamma_g^2$ over the accelerating phase

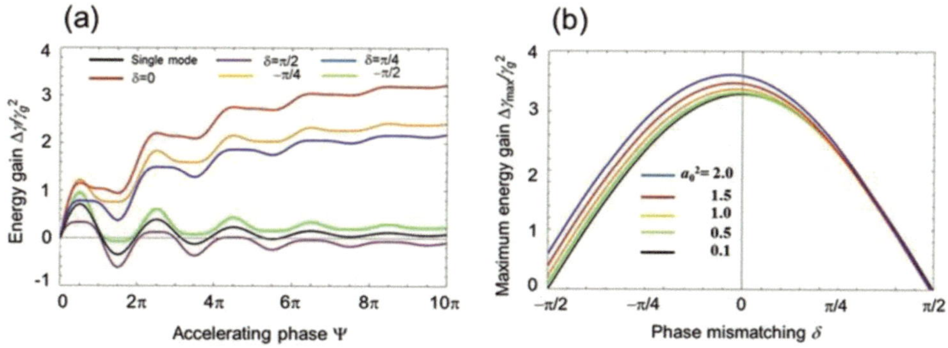

Figure 2.
(a) The evolution of the energy gain normalized to γ_g^2 of the mode mixing LPA comprising two hybrid modes EH_{11} and EH_{12} coupled to the airy beam intensity of $a_0^2 = 1$ with the coupling efficiency $C_1 = 0.4022$ and $C_2 = 0.4986$, respectively, as a parameter of the mismatching phase δ. The black curve shows that for the single-mode LPA with $C_1 = 1$ and $C_2 = 0$. (b) The maximum attainable energy gain of the two-mode mixing LPA for various normalized intensities a_0^2 as a function of the phase mismatching.

$0 \leq \Psi \leq \pi/2$, the two-mode mixing LPA with the phase matching, i.e., $\delta = 0$, is attainable to the maximum energy gain $\Delta\gamma_{\max} = 3.2\gamma_g^2$ over the accelerating phase region $\Delta\Psi = 10\pi$, as shown in **Figure 2**. It is noted that significant enhancement of the energy gain is attributed to a large energy transfer efficiency from the laser pulse to the wakefield, i.e., $\eta_{\text{laser}\to\text{wake}} \sim 96\%$ over the accelerating phase region $\Delta\Psi = 10\pi$, while the energy transfer efficiency for the single-mode LPA is $\eta_{\text{laser}\to\text{wake}} \sim 17\%$ over the accelerating phase region $\Delta\Psi = \pi/2$.

The average energy gain of electrons contained in the bunch with the root-mean-square (rms) bunch length $k_p\sigma_z$ and longitudinal Gaussian density distribution $\rho_\parallel(\Psi') = e^{-\Psi'^2/2k_p^2\sigma_z^2}/(\sqrt{2\pi}k_p\sigma_z)$ can be estimated as $\langle\Delta\gamma\rangle = \langle G(\Psi)\rangle - \langle G(\Psi_0)\rangle$, where

$$\langle G(\Psi)\rangle = \int_{-\infty}^{\infty} \rho_\parallel(\Psi' - \Psi)G(\Psi')d\Psi' \approx \sqrt{\frac{\pi}{2}}a_0{}^2\gamma_g^2(C_1 + C_2)e^{-\left(1+4\alpha_d a_0^2\Psi\right)/2}$$
$$\times\left[\frac{e^{-k_p^2\sigma_z^2/2}\left(\sin\Psi - 2\alpha_d a_0^2\cos\Psi\right)}{1 + 4\alpha_d^2 a_0^4} - \frac{\sqrt{C_1C_2}}{2(C_1 + C_2)}\left(\frac{1}{\alpha_d a_0^2} - \frac{e^{-2k_p^2\sigma_z^2}\left(\sin 2\Psi - \alpha_d a_0^2\cos 2\Psi\right)}{1 + \alpha_d^2 a_0^4}\right)\right]. \tag{17}$$

Figure 3 shows the evolution of the energy gain and the maximum attainable energy gain averaged over electrons in a Gaussian bunch with various rms lengths. It is noted that the maximum attainable energy gain at $\Psi \to \infty$ exhibits weak dependence on the initial bunch phase Ψ_0 for a long bunch and that the minimum energy spread occurs at $\Psi_0 \sim 0$ for different bunch lengths.

3.2 Beam loading

In the linear regime, a solution of the Green's function for the beam-driven wakefield excited by a charge bunch with bi-Gaussian density distribution $\rho_b = \rho_\parallel(\xi)\rho_\perp(r)$, i.e., $\rho_\parallel(\xi) = qn_b\exp\left(-\xi^2/2\sigma_z^2\right)$ and $\rho_\perp(r) = \exp\left(-r^2/2\sigma_r^2\right)$ for the rms bunch length σ_z, rms bunch radius σ_r, and particle charge q ($+e$ for a positron beam and $-e$ for an electron beam), is written as $E_{zb}(r,\xi) = Z(\xi)R(r)$, where $\xi = z - ct$ is the coordinate in the co-moving frame of a relativistic electron beam with $v_z \simeq c$ and r is the radial, transverse coordinate of an electron beam having a cylindrical symmetry [38]. Here, the longitudinal and transverse plasma responses are obtained as

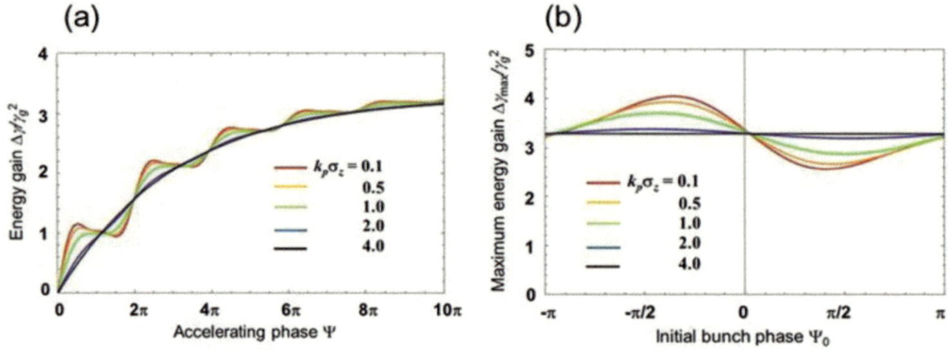

Figure 3.
*(a) The evolution of the energy gain $\Delta\gamma/\gamma_g^2$ of the mode mixing LPA driven by two hybrid modes EH_{11} and EH_{12} with the same parameters as those of **Figure 2a** for various rms bunch lengths $k_p\sigma_z$. (b) The maximum attainable energy gain of two-mode mixing LPA for various rms bunch lengths as a function of the initial phase of the bunch center with respect to the maximum accelerating field.*

$$Z(\xi) = -4\pi\int_{\xi}^{\infty} d\xi' \rho_{\parallel}(\xi')\cos k_p(\xi-\xi') = -(2\pi)^{3/2} q n_b \sigma_z e^{-k_p^2\sigma_z^2/2} \times\left[\cos k_p\xi\left(1-\Re\mathrm{erf}\frac{\xi/\sigma_z + ik_p\sigma_z}{\sqrt{2}}\right) + \sin k_p\xi\,\Im\mathrm{erf}\frac{\xi/\sigma_z + ik_p\sigma_z}{\sqrt{2}}\right], \tag{18}$$

and inside the bunch ($r < r'$)

$$R(r) = \left(k_p^2/2\pi\right)\int_0^{2\pi} d\theta \int_0^{\infty} r' dr' \rho_{\perp}(r') K_0\left(k_p\left|\vec{r}-\vec{r'}\right|\right) = \left(k_p^2\sigma_r^2/2\right) e^{k_p^2\sigma_r^2/2}\Gamma\left(0, k_p^2\sigma_r^2/2\right) J_0(k_p r), \tag{19}$$

where K_0 is the modified Bessel function of the second kind and $\Gamma(\alpha,x) = \int_x^{\infty} e^{-t}t^{\alpha-1}dt$ is the incomplete Gamma function of the second kind. Combining the longitudinal and transverse solutions, the wakefield excited by a bi-Gaussian-shaped bunch is obtained as

$$E_{zb}(r,\xi)/E_0 = -(q/e)k_p r_e N_b \Theta_G(\sigma_r,\sigma_z) J_0(k_p r) \times\left[\cos k_p\xi\left(1-\Re\mathrm{erf}\frac{\xi/\sigma_z + ik_p\sigma_z}{\sqrt{2}}\right) + \sin k_p\xi\,\Im\mathrm{erf}\frac{\xi/\sigma_z + ik_p\sigma_z}{\sqrt{2}}\right], \tag{20}$$

where $r_e = e^2/m_e c^2$ is the electron classical radius, $N_b = (2\pi)^{3/2}\sigma_r^2\sigma_z n_b$ the number of particles in a bunch, and $\Theta_G(\sigma_r,\sigma_z) \equiv e^{-k_p^2\sigma_z^2/2} e^{k_p^2\sigma_r^2/2}\Gamma\left(0, k_p^2\sigma_r^2/2\right)$. If we consider a laser-driven wakefield E_{zL} excited by two mixing hybrid modes accelerating the electron beam in a gas-filled capillary, the net longitudinal electric field, i.e., the beam loading field, experienced by the electron beam is given by $E_{z\mathrm{BL}} = E_{zL} + E_{zb}$. From Eqs. (13) and (20), the beam loading field at $r = 0$ consisting of the laser- and beam-driven wake, where the electron bunch is located at $\Psi = \Psi_b$ in the laser co-moving frame, i.e., $k_p\xi \approx \Psi - \Psi_b$, yields

$$E_{z\mathrm{BL}}(0,\Psi)/E_0 = \hat{E}_{z\mathrm{L}}(\Psi)\cos\Psi + k_p r_e N_b \Theta_G(\sigma_r,\sigma_z) \times\left[\cos(\Psi-\Psi_b)\Re\mathrm{erfc}\left(\frac{\Psi-\Psi_b}{\sqrt{2}k_p\sigma_z} + i\frac{k_p\sigma_z}{\sqrt{2}}\right) + \sin(\Psi-\Psi_b)\Im\mathrm{erf}\left(\frac{\Psi-\Psi_b}{\sqrt{2}k_p\sigma_z} + i\frac{k_p\sigma_z}{\sqrt{2}}\right)\right], \tag{21}$$

where $\hat{E}_{zL}(\Psi) = -\sqrt{\pi/8}a_0{}^2e^{-(1+4\alpha_d a_0^2\Psi)/2}(C_1 + C_2 + \sqrt{C_1C_2}\cos\Psi)$. A loss of the energy gain due to the beam wakefield $E_{zb} = k_p r_e N_b \Theta_G(\sigma_r, \sigma_z)$ at the bunch center is

$$\Delta\gamma_{BL} = -\left(2\gamma_g^2/E_0\right)\int_{\Psi_0}^{\Psi} E_{zb}(\Psi')d\Psi' = -2\gamma_g^2\Theta_G(\sigma_r, \sigma_z)(\Psi - \Psi_0), \tag{22}$$

and the rms energy spread due to the beam loading is estimated as

$$\sigma_{\Delta\gamma_{BL}} \approx 2^{1/4}\gamma_g^2 k_p r_e N_b e^{k_p^2\sigma_r^2/2}\Gamma\left(0, k_p^2\sigma_r^2/2\right)S(k_p\sigma_z)(\Psi - \Psi_0), \tag{23}$$

where $S(k_p\sigma_z) = \left|(2/\pi)^{1/2} - k_p\sigma_z e^{-k_p^2\sigma_z^2/2}\text{erf}\left(k_p\sigma_z/\sqrt{2}\right)\right|$ has the minimum $S = 0.35$ at $k_p\sigma_z = 1.26$.

3.3 Betatron motion

In the wakefield, an electron moving along the z-axis undergoes a transverse focusing force $F_\perp^B = -m_e c^2 F_r x/r$ at the transverse displacement x and exhibits the betatron motion. Taking into account $F_r \approx (\partial F_r/\partial r)r$ near the z-axis $r \approx 0$, the focusing force is written by $F_\perp^B/m_e c^2 = -(\partial F_r/\partial r)x = -K^2 x$, where $K = (\partial F_r/\partial r)^{1/2}$ is the focusing constant. For the optimum pulse length of $k_p c\tau = \sqrt{2}$ and $\Psi = k_p z/\left(2\gamma_g^2\right) \gg k_p c\tau/\left(2\gamma_g^2\right) = 1/\left(\sqrt{2}\gamma_g^2\right)$ in Eq. (8), transverse laser wakefield in the matching condition $\Delta k_{z12} \approx \left(u_2^2 - u_1^2\right)/\left(k_0 R_c^2\right) = k_p/\gamma_g^2$ is given by

$$\begin{aligned}\frac{F_{rL}}{E_0} = \sqrt{\frac{\pi}{2}}\frac{a_0^2 e^{-1/2}}{\sqrt{\gamma_g\left(u_2^2 - u_1^2\right)}}&\left[\left(C_1 u_1 J_0\left(\frac{u_1 r}{R_c}\right)J_1\left(\frac{u_1 r}{R_c}\right) + C_2 u_2 J_0\left(\frac{u_2 r}{R_c}\right)J_1\left(\frac{u_2 r}{R_c}\right)\right)\sin\Psi\right.\\ &\left.+\frac{\sqrt{C_1C_2}}{2}\left(u_1 J_0\left(\frac{u_2 r}{R_c}\right)J_1\left(\frac{u_1 r}{R_c}\right) + u_2 J_0\left(\frac{u_1 r}{R_c}\right)J_1\left(\frac{u_2 r}{R_c}\right)\right)\sin 2\Psi\right].\end{aligned} \tag{24}$$

Transverse wakefield excited by the electron bunch is obtained from Eq. (20) according to the Panofsky-Wenzel theorem [39], $\partial E_z/\partial r = \partial(E_r - B_\theta)/\partial\xi$, leading to the beam focusing force [40].

$$\begin{aligned}F_{rb}(r,\xi)/E_0 &= (E_r - B_\theta)/E_0 = (q/e)k_p r_e N_b \Theta_G(\sigma_r, \sigma_z)J_1(k_p r)\\ &\times\left[\sin k_p\xi\left(1 - \Re\text{erf}\frac{\xi/\sigma_z + ik_p\sigma_z}{\sqrt{2}}\right) - \cos k_p\xi\,\Im\text{erf}\frac{\xi/\sigma_z + ik_p\sigma_z}{\sqrt{2}}\right].\end{aligned} \tag{25}$$

At the bunch center $\xi = 0$, the on-axis beam focusing strength at $r = 0$

$$\frac{\partial F_{rb}}{E_0 k_p \partial r} = \frac{k_p r_e N_b}{\sqrt{\pi}}e^{k_p^2\sigma_r^2/2}\Gamma\left(0, \frac{k_p^2\sigma_r^2}{2}\right)\text{F}\left(\frac{k_p\sigma_z}{\sqrt{2}}\right), \tag{26}$$

where $\text{erf}\left(ik_p\sigma_z/\sqrt{2}\right) = \left(2i/\sqrt{\pi}\right)\int_0^{k_p\sigma_z/\sqrt{2}}e^{z^2}dz = \left(2i/\sqrt{\pi}\right)e^{\frac{k_p{}^2\sigma_z^2}{2}}\text{F}\left(k_p\sigma_z/\sqrt{2}\right)$ and $\text{F}(x) = e^{-x^2}\int_0^x e^{z^2}dz$ is Dawson's integral. Near-axis electrons experience the normalized accelerating and focusing gradients at $r = 0$, as obtained from Eqs. (21), (24), and (26)

$$\frac{E_{zBL}}{E_0} = -\frac{1}{2}\sqrt{\frac{\pi}{2}}a_0^2 e^{-(1+4\alpha a_0^2\Psi)/2}(C_1 + C_2)\left[\cos\Psi + \frac{\sqrt{C_1C_2}}{C_1 + C_2}(1 + \cos 2\Psi)\right] + k_p r_e N_b \Theta_G(\sigma_r, \sigma_z), \tag{27}$$

and

$$\frac{\partial F_{rBL}}{E_0 k_p \partial r} = -\frac{1}{2}\sqrt{\frac{\pi}{2}}a_0^2 e^{-(1+4\alpha a_0^2\Psi)/2}\frac{u_1^2 C_1 + u_2^2 C_2}{\gamma_g(u_2^2 - u_1^2)}\left[\sin\Psi + \frac{(u_1^2 + u_2^2)\sqrt{C_1C_2}}{2(u_1^2 C_1 + u_2^2 C_2)}\sin 2\Psi\right] + \frac{k_p r_e N_b}{\sqrt{\pi}}\Theta_F(\sigma_r, \sigma_z), \tag{28}$$

where $\Theta_F(\sigma_r, \sigma_z) \equiv e^{k_p^2\sigma_r^2/2}\Gamma\left(0, k_p^2\sigma_r^2/2\right)F\left(k_p\sigma_z/\sqrt{2}\right)$ is the bunch form factor for a bi-Gaussian profile with the rms bunch radius σ_r and length σ_z.

The equations of motion of an electron propagating in the wakefield behind the laser pulse is written as [41].

$$\frac{d^2\bar{x}}{d\bar{t}^2} - \frac{\overline{E}_z}{\gamma}\frac{d\bar{x}}{d\bar{t}} + \frac{\overline{K}^2}{\gamma}\bar{x} = 0,\ \frac{d\gamma}{d\bar{t}} = -\overline{E}_z, \tag{29}$$

where $\bar{x} = k_p x$ and $\bar{t} = \omega_p t$ are the normalized variables of x and t, respectively. Here the longitudinal wakefield and focusing constant at $r = 0$ are defined as $\overline{E}_z \equiv E_{zBL}/E_0$ and $\overline{K}^2 \equiv (1/E_0 k_p)(\partial F_{rBL}/\partial r)$, respectively. If one can assume that $\overline{E}_z$ and $\overline{K}$ are constant along the particle trajectory, introducing a new variable $s = \left(4\gamma\overline{K}^2/\overline{E}_z^2\right)^{1/2}$ to obtain the differential equation $s\left(d^2\bar{x}/ds^2\right) + (d\bar{x}/ds) + s\bar{x} = 0$, general solutions of which are the Bessel functions of the first kind $J_0(s)$ and the second kind $Y_0(s)$, the solutions of the coupled equations are given by Eqs. (14) and (15) for γ, and the transverse position and velocity [41].

$$\begin{pmatrix} \bar{x}(s) \\ \beta_x(s) \end{pmatrix} = \mathbf{M}(s|s_0)\begin{pmatrix} \bar{x}_0(s_0) \\ \beta_{x0}(s_0) \end{pmatrix}, \tag{30}$$

where $\beta_x = \beta_g(d\bar{x}/d\bar{t})$, subscripts "0" denote the initial values, and

$$\mathbf{M}(s|s_0) = \begin{pmatrix} \frac{\pi}{2}s_0[J_1(s_0)Y_0(s) - Y_1(s_0)J_0(s)] & \frac{\pi\gamma_0}{E_{z0}}[J_0(s)Y_0(s_0) - Y_0(s)J_0(s_0)] \\ -\frac{\pi E_z s_0 s}{4\gamma}[J_1(s)Y_1(s_0) - Y_1(s)J_1(s_0)] & \frac{\pi E_z}{2E_{z0}}\frac{\gamma_0}{\gamma}s[J_1(s)Y_0(s_0) - Y_1(s)J_0(s_0)] \end{pmatrix}. \tag{31}$$

While the electron stays in the focusing region of the wakefield, i.e., $\partial F_r/\partial r > 0$, the electron exhibits betatron oscillation at the frequency given by $\omega_\beta = ds/dt = \omega_p\overline{K}/\gamma^{1/2}$. Contrarily, when the electron moves to the defocusing region where $\partial F_r/\partial r < 0$ and s becomes imaginary, the amplitude of the electron trajectory increases monotonically as a result of the Bessel functions being transformed to the modified Bessel functions, leading to ejection of the electron from the wakefield [41]. Hence, the requirement of betatron oscillation in the focusing region $\overline{K}^2 > 0$

demands that the minimum number of electrons contained in a bunch should be injected into the plasma as given by

$$N_b \geq \frac{\sqrt{\pi}}{k_p r_e \Theta_F(\sigma_r, \sigma_z)} \left| \frac{\partial F_{rL}(\Psi)}{E_0 k_p \partial r} \right|_{\max}, \tag{32}$$

for a bi-Gaussian bunch with the rms radius σ_r and length σ_z. **Figure 4** shows a map of the bunch form factor $\Theta_F(\sigma_r, \sigma_z)$ and the minimum number of electrons contained in a bunch requisite for the beam self-focusing strength larger than the defocusing strength in the laser-driven wakefield for the EH_{11}-EH_{12} mode mixing LPA. It is noted that the minimum value of the requisite electron number occurs at the bunch length $k_p\sigma_z = 1.31$ for various bunch radii, e.g., $N_b \geq 3.63 \times 10^7$ for $k_p\sigma_r = 1$ and $N_b \geq 7.05 \times 10^6$ for $k_p\sigma_r = 0.1$, as shown in **Figure 4**.

In the bunch containing the requisite number of particles, an electron undergoes betatron motion throughout the whole accelerating phase, as shown in **Figure 5**, where the trajectory and momentum of the electron in the bunch with the number of electrons $N_b = 1 \times 10^8$ and length $k_p\sigma_z = 1.3$ are calculated from Eq. (30) in 10^5 segments of the laser wakefield phase excited in the plasma with density $n_e = 1 \times 10^{18}$ cm^{-3}. Note that the betatron oscillation exhibits beats with the amplitude modulation due to the accelerating wakefield.

3.4 Effects of radiation reaction and multiple Coulomb scattering

A beam electron propagating in the wakefield undergoes betatron motion that induces synchrotron (betatron) radiation at high energies. The synchrotron radiation causes the radiation damping of particles and affects the energy spread and transverse emittance via the radiation reaction force. Furthermore, a notable difference of plasma-based accelerators from vacuum-based accelerators is the presence of the multiple Coulomb scattering between beam electrons and plasma ions, which counteracts the beam focusing due to the transverse wakefield and radiation damping due to betatron radiation. The comprehensive motion of an electron traveling along the z-axis is described as

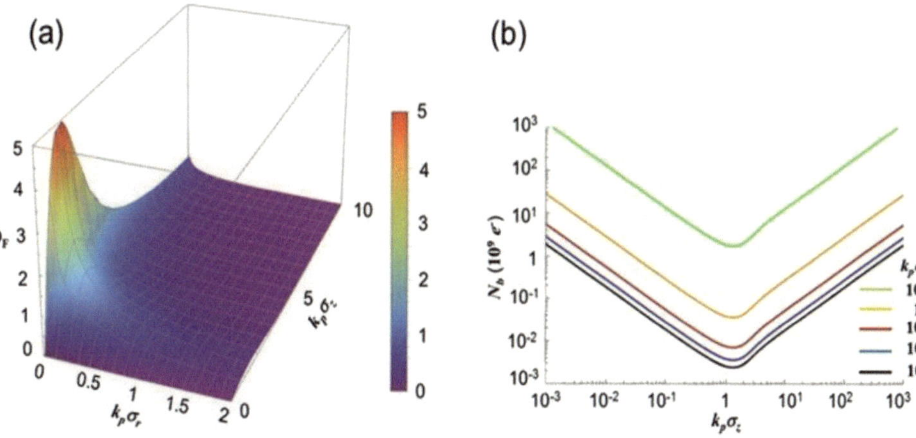

Figure 4.
(a) A map of the bunch form factor $\Theta_F(\sigma_r, \sigma_z)$ for the beam self-focusing strength of a bi-Gaussian bunch as a function of the dimensionless rms bunch radius $k_p\sigma_r$ and length $k_p\sigma_z$. (b) The minimum number of electrons contained in a Gaussian bunch requisite for the beam self-focusing strength larger than the defocusing strength from the laser-driven wakefield in the EH_{11}-EH_{12} mode mixing LPA.

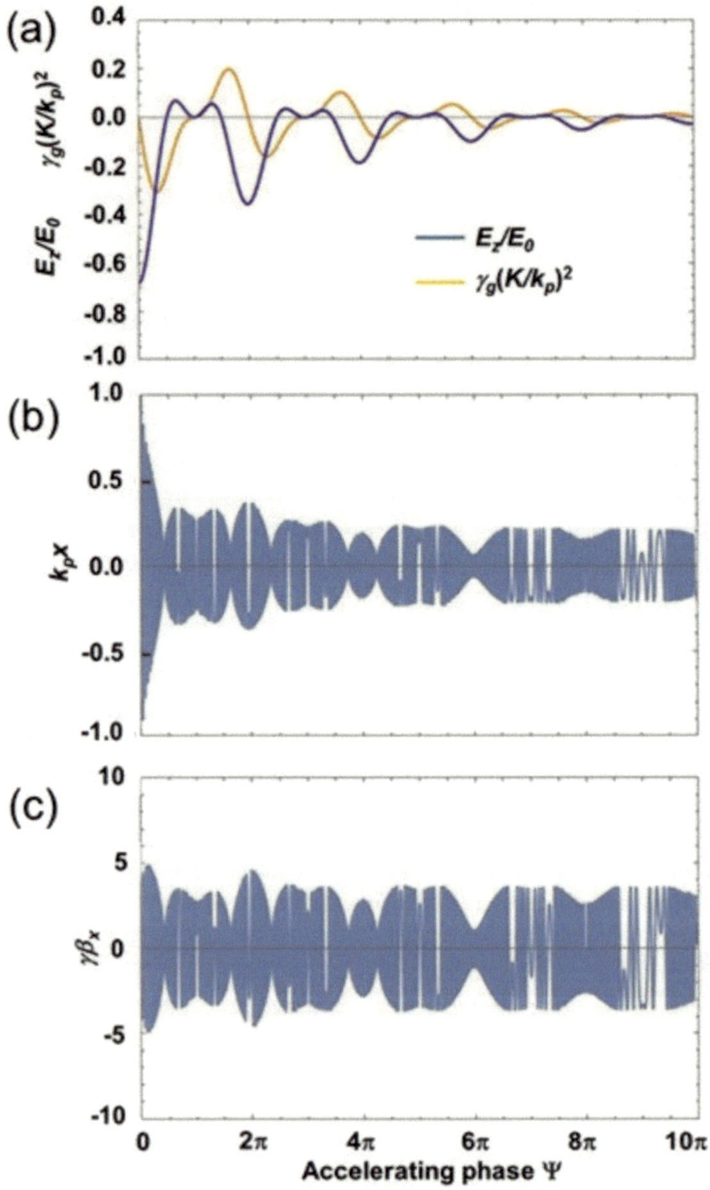

Figure 5.
(a) Normalized accelerating wakefield E_z/E_0 and focusing strength $\gamma_g K^2/k_p^2 = \gamma_g(1/E_0 k_p)(\partial F_{rL}/\partial r)$ in the EH_{11}-EH_{12} mode mixing LPA with the same parameters as those of ***Figure 2a*** *as a function of the accelerating phase Ψ. (b and c) normalized trajectory $k_p x$ and transverse momentum $\gamma\beta_x$ of the electron with the initial values $m_e c^2\gamma_0 = 100\,\mathrm{MeV}$, $k_p x_0 = 1$, and $\gamma\beta_{x0} = 0$ in the bunch with the number of electrons $N_b = 1\times 10^8$ and length $k_p\sigma_z = 1.3$ for betatron motion in the laser wakefields (a) with electron density $n_e = 1\times 10^{18}$ cm^{-3}.*

$$\frac{du_x}{cdt} = \frac{F_\perp^{\mathrm{B}}}{m_e c^2} + \frac{F_x^{\mathrm{R}}}{m_e c^2} + \frac{du_x^{\mathrm{S}}}{cdt}, \; \frac{du_z}{cdt} = -k_p\frac{E_z}{E_0} + \frac{F_z^{\mathrm{R}}}{m_e c^2}, \tag{33}$$

where $\mathbf{u} = \mathbf{p}/m_e c$ is the normalized electron momentum, $\mathbf{F}^{\mathrm{R}}$ the radiation reaction force, and $u_x^{\mathrm{S}} \approx \gamma\theta_x$ the transverse kick in momentum projected onto the x-plane due to multiple Coulomb scattering through small deflection angles θ.

For the classical expression of the radiation reaction force given by [42].

$$\frac{\mathbf{F}^{\mathrm{R}}}{m_e c\tau_R} = \frac{d}{dt}\left(\gamma\frac{d\mathbf{u}}{dt}\right) + \gamma\mathbf{u}\left[\left(\frac{d\gamma}{dt}\right)^2 - \left(\frac{d\mathbf{u}}{dt}\right)^2\right], \tag{34}$$

where $\gamma = (1 + u^2)$ is the relativistic Lorentz factor of the electron and $\tau_R = 2r_e/3c \simeq 6.26 \times 10^{-24}$s, assuming $u_z \gg u_x$ and $dx/dt = cu_x/\gamma \simeq cu_x/u_z$, the radiation reaction force Eq. (34) is approximately read as [43].

$$F_x^{\mathrm{R}}/(m_e c^2) \simeq -c\tau_R K^2 u_x(1 + K^2\gamma x^2), F_z^{\mathrm{R}}/(m_e c^2) \simeq -c\tau_R K^4\gamma^2 x^2. \tag{35}$$

Since the scale length of the radiation reaction, i.e., $c\tau_R = 2r_e/3 \simeq 1.9$ fm, is much smaller than that of the betatron motion, i.e., $\sim \lambda_p\sqrt{\gamma}$, the radiation reaction force is considered as a perturbation in the betatron motion.

A beam electron of the incident momentum $p = \gamma m_e v$, passing a nucleus of charge Ze at impact parameter b in the plasma, suffers an angular deflection $\theta = \Delta p/p \simeq 2e^2 Z/(pbv)$ due to Coulomb scattering [44]. The successive collisions of the relativistic beam electrons with $v \sim c$ while traversing the plasma of ion density $n_i = n_e/Z$ results in an increase of the mean square deflection angle at a rate [8, 44].

$$\frac{d\langle\theta^2\rangle}{cdt} = \frac{8\pi n_i Z^2 r_e^2}{\gamma^2}\ln\left(\frac{b_{\max}}{b_{\min}}\right) = \frac{2k_p^2 r_e Z}{\gamma^2}\ln\left(\frac{\lambda_D}{R_N}\right), \tag{36}$$

where $b_{\max} = \lambda_D = (T_e/4\pi n_e e^2)^{1/2}$ is the plasma Debye length at the temperature T_e and $R_N \approx 1.4A^{1/3}$ fm is the effective Coulomb radius of the nucleus with the mass number A. Here, the logarithm $\ln(b_{\max}/b_{\min})$ is approximated as $\ln(\lambda_D/R_N) \approx 24.7\left[1 + 0.047\log\left(n_e T_e A^{2/3}\right)\right]$ for n_e $[10^{16}\ \mathrm{cm}^{-3}]$ and T_e [eV] [45]. The multiple-scattering distribution for the projected angle θ_x is approximately Gaussian for small deflection angles, given by the probability distribution function $P(\theta_x) = 1/(\pi\langle\theta^2\rangle)^{1/2}\exp(-\theta_x^2/\langle\theta^2\rangle)$. Thus, the transverse momentum $u_x^{\mathrm{S}} \approx \gamma\theta_x$ is obtained from using the normal distribution with the standard deviation $(\langle\theta^2\rangle/2)^{1/2}$ around the mean angle 0 at the successive time step along the particle trajectory.

The electron orbit and energy are obtained from the solutions of the coupled equations in Eq. (33) describing the single particle motion in the segmented phase, where $\overline{E}_z$ and $\overline{K}$ are assumed to be constant over the phase advance $\Delta\Psi$. Provided the initial values of $\overline{x}_0$ and β_{x0} are specified from the energy γ_0, relative energy spread $\Delta\gamma/\gamma_0$, and normalized emittance ε_{n0} of the injected beam, $\gamma(s)$, $\overline{x}(s)$, and $\beta_x(s)$ are first calculated from Eqs. (14) and (30) using $s(\Psi)$, where $\Psi = \Psi_0 + \Delta\Psi$ is the phase at next step. Thus, the effects of the radiation reaction and multiple Coulomb scattering are obtained as follows:

$$\beta_x(s) = \beta_x^{\mathrm{B}}(s) + \Delta\beta_x^{\mathrm{R}}(s_0) + \Delta\beta_x^{\mathrm{S}}(s_0),\ \gamma(\Psi) = \gamma^{\mathrm{A}}(\Psi) + \Delta\gamma^{\mathrm{R}}(\Psi_0), \tag{37}$$

where $\beta_x^{\mathrm{B}}(s)$ and $\gamma^{\mathrm{A}}(\Psi)$ are the solutions obtained from Eqs. (30) and (14), respectively; $\Delta\beta_x^{\mathrm{R}}(s_0)$ and $\Delta\gamma^{\mathrm{R}}(\Psi_0)$ are correction terms for the effect of the radiation reaction force, given by

$$\Delta\beta_x^{\mathrm{R}} = -2C_R\gamma_g\beta_{x0}\overline{K}_0^2\left(1 + \gamma_0\overline{K}_0^2\overline{x}_0^2\right)\Delta\Psi,\ \Delta\gamma^{\mathrm{R}}(\Psi_0) = -2C_R\gamma_g\gamma_0^2\overline{K}_0^2\overline{x}_0^2\Delta\Psi, \tag{38}$$

with $C_R = k_p c \tau_R \gamma_g = (2/3) k_p r_e \gamma_g = 11.8 \times 10^{-9} [\mu\text{m}]/\lambda_0$ and $\overline{K}_0^2 = \overline{K}^2(\Psi_0)$; and $\Delta\beta_x^S(s_0) = \theta_x$ is a projected angle due to multiple Coulomb scattering, the standard deviation of which is obtained from Eq. (36) for $\lambda_0 = 1\ \mu$m as

$$\sigma_{\theta_x} = \sqrt{\langle\theta^2\rangle/2} \approx 2.66 \times 10^{-4} \left[\gamma_g \Delta\Psi \ln(\lambda_D/R_N)\right]^{1/2} / \gamma_0. \tag{39}$$

The radiated power of the electron in the classical limit is given by [42, 43].

$$P_{\text{rad}} = \frac{2e^2\gamma^2}{3c}\left[\left(\frac{d\mathbf{u}}{dt}\right)^2 - \left(\frac{d\gamma}{dt}\right)^2\right] = \frac{2e^2\gamma^2}{3m_e^2c^3}\left[|\mathbf{F}_{\text{ext}}|^2 - |\mathbf{F}_{\text{ext}}\cdot\mathbf{u}/\gamma|^2\right]. \tag{40}$$

where $\mathbf{F}_{\text{ext}}$ is the external force and $m_e c\, d\gamma/dt = \mathbf{F}_{\text{ext}}\cdot\mathbf{u}/\gamma$ is used. For a relativistic electron with $u_x^2 \ll \gamma^2$ and $u_z \sim \gamma$, taking into account $\mathbf{F}_{\text{ext}} = F_\perp \mathbf{e}_x + F_\parallel \mathbf{e}_z$ with $F_\perp = -m_e c^2 K^2 x$ and $F_\parallel = -eE_z$, the radiated power can be written as $P_{\text{rad}} = 2e^2\gamma^2 {F_\perp}^2/(3m^2c^3) = mc^4 \tau_R \gamma^2 K^4 x^2$, which means the radiative damping rate $\nu_R = P_{\text{rad}}/(\gamma m_e c^2) = \tau_R c^2 \gamma K^4 x^2$. Thus, a total radiation energy loss along the particle orbit is estimated as

$$\Delta\gamma_{\text{rad}} = \frac{1}{m_e c^2}\int_{t_0}^{t} dt P_{\text{rad}}(t) = \sum\left|\Delta\gamma^{\text{R}}(\Psi_0)\right|. \tag{41}$$

3.5 Numerical studies of the single-particle dynamics in a single stage

Numerical calculations of the single-particle dynamics can be carried out throughout the segments in phase Ψ for a set of test particles under the initial conditions, and then the underlying beam parameters can be obtained as an ensemble average over test particles: for instance, the mean energy is calculated as $\langle\gamma\rangle = \sum_i \gamma_i / N_p$, where γ_i is the energy of the i-th particle and N_p the number of test particles, and the energy spread is defined as $\sigma_\gamma = \left(\langle\gamma^2\rangle - \langle\gamma\rangle^2\right)^{1/2}$. The normalized transverse emittance is obtained from

$$\varepsilon_n = \left[\left\langle(x - \langle x\rangle)^2\right\rangle\left\langle(u_x - \langle u_x\rangle)^2\right\rangle - \langle(x - \langle x\rangle)(u_x - \langle u_x\rangle)\rangle^2\right]^{1/2}, \tag{42}$$

where $u_x = \gamma\beta_x$ is the dimensionless transverse momentum.

The particle orbit and energy can be numerically tracked by using the solutions of the single particle motion (Eqs. (30) and (14)) associated with the perturbation arising from the effects of the radiation reaction and multiple Coulomb scattering, as given by Eqs. (38) and (39), respectively. The simulation of particle tracking can be carried out by using an ensemble of 10^4 test particles, for which the initial values at the injection and the deflection angles due to the multiple Coulomb scattering at each segment in a phase step $\Delta\Psi/400$, where $\Delta\Psi = 10\pi$ is the phase advance in the single stage, are obtained from the random number generator for the normal distribution, assuming that the particle beam with the rms bunch length $\sigma_z = 16\ \mu$m ($k_p\sigma_z = 3$) containing $N_b = 1 \times 10^8$ electrons (16 pC) is injected into the capillary accelerator operated at the plasma density of $n_e = 1 \times 10^{18}$ cm^{-3} from the external injector at the injection energy $E_{\text{inj}} = m_e c^2 \gamma_0$ and the initial normalized transverse emittance ε_{n0} in the condition initially matched to laser wakefields, namely, the initial bunch radius $\overline{x}_0 = k_p\sigma_0 = (k_p\varepsilon_{n0})^{1/2} / \left(\gamma_0 \overline{K}^2\right)^{1/4}$ and momentum

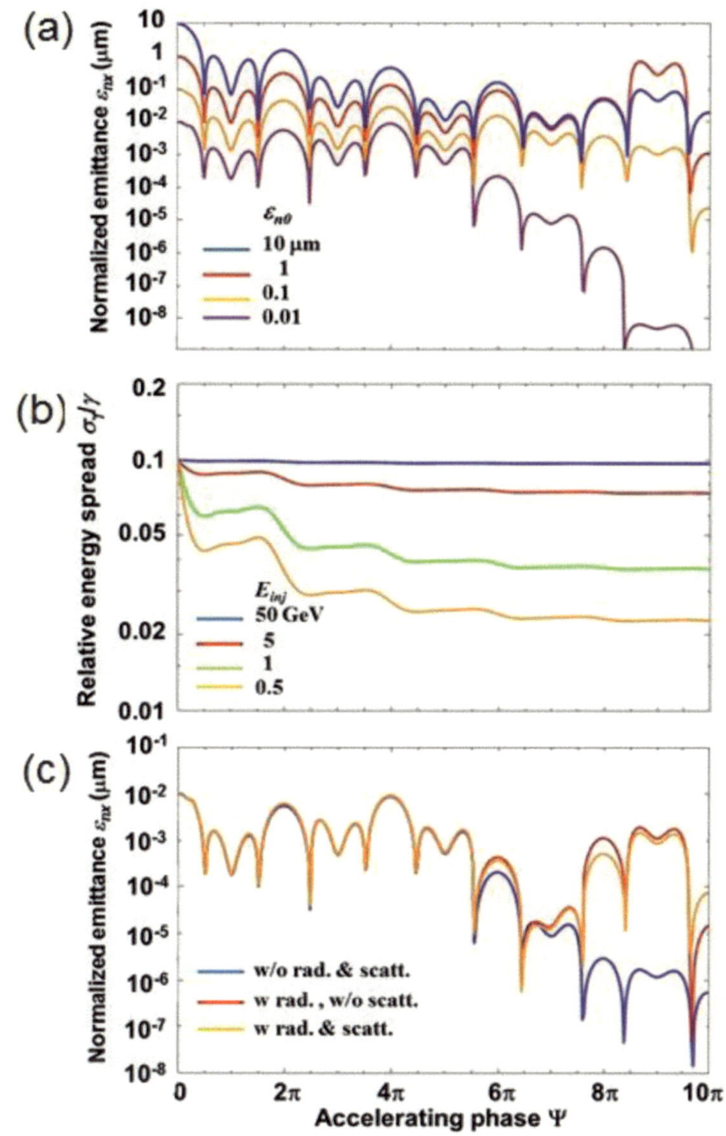

Figure 6.
*Numerical results of the beam dynamics study on the two-mode mixing single-stage LPA (**Figure 1a**) at the plasma density $n_e = 1 \times 10^{18}$ cm^{-3} and number of electrons $N_b = 1 \times 10^8$ in a bunch with length $k_p\sigma_z = 3$. (a) Evolution of transverse normalized emittance ε_{nx} from various initial values for the initial energy $E_{inj} = m_ec^2\gamma_0 = 1$ GeV and relative energy spread $\sigma_\gamma/\gamma_0 = 0.1$ without radiation and multiple Coulomb scattering. (b) Evolution of relative energy spread σ_γ/γ from the initial value of 0.1 for various initial energies E_{inj} due to radiation reaction without multiple Coulomb scattering. (c) Evolution of transverse normalized emittance from the initial value $\varepsilon_{n0} = 0.01$ μm for various cases with and without the radiation reaction and multiple Coulomb scattering.*

$\gamma_0\beta_0 = \left(\gamma_0\overline{K}^2\right)^{1/4}\left(k_p\varepsilon_{n0}\right)^{1/2}$ with the focusing strength $\overline{K}^2$, given by Eq. (28). **Figure 6a** and **b** show the results of simulations for the evolution of transverse normalized emittance ε_{nx} from various initial values ε_{n0} at the initial phase $\Psi_0 = 0$ and that of the relative energy spread σ_γ/γ from the initial spread of $\sigma_\gamma/\gamma_0 = 0.1$ for various initial energies due to the effect of the radiation reaction, respectively. The effect of the multiple Coulomb scattering is shown in **Figure 6c**, indicating a significant growth of the normalized emittance in the latter half of the stage. In this simulation, the multiple Coulomb scattering has been considered for a helium plasma with $A = 4$, $Z = 2$, and $T_e = 100$ eV. Since the normalized emittance, defined by Eq. (42), is approximately calculated as $\varepsilon_{nx} \sim |\delta x||\delta(\gamma\beta_x)|$, where δx and $\delta(\gamma\beta_x)$ are the amplitudes of the transverse displacement and dimensionless momentum, the evolution of the normalized emittance traces the envelope of the betatron oscillation of the single particle, as seen in **Figures 5** and **6**. Note that the electron motion of coupled equation in Eq. (29) includes the nonlinear damping term $-\left(\overline{E}_z/\gamma\right)\left(d\overline{x}/d\overline{t}\right)$, which induces the amplitude decrease in the electron acceleration phase, while the betatron motion of the electron undergoing only a linear focusing force with a constant $\overline{K}$ is described by a simple harmonic oscillator at a constant energy $m_e c^2\gamma$, i.e., no acceleration field $\overline{E}_z = 0$, forming the constant envelope of the betatron amplitude for the matched condition of bunch size $\sigma_x^2 = \varepsilon_{nx}/\left(\gamma\overline{K}\right)^{1/2}$, for which the normalized emittance is conserved.

4. Beam dynamics in multistage two-mode mixing LPAs

4.1 Seamless stage coupling with a variable curvature plasma channel

A gas-filled capillary waveguide made of metallic or dielectric materials can make it possible to comprise a seamless staging without the coupling section, where a fresh laser pulse and accelerated particle beam from the previous stage are injected so as to minimize coupling loss in both laser and particle beams and the emittance growth of particle beams due to the mismatch between the injected beam and plasma wakefield. For dephasing limited laser wakefield accelerators, the total linac length will be minimized by choosing the coupling distance to be equal to a half of the dephasing length [9]. A side coupling of laser pulse through a curved capillary waveguide [46–48] diminishes the beam-matching section consisting of a vacuum drift space and focusing magnet beamline [9]. Furthermore, the proposed scheme comprising seamless capillary waveguides can provide us with suppression of synchrotron radiation from high-energy electron (positron) beams generated by betatron oscillation in plasma-focusing channels and delivery of remarkably small normalized emittance from the linac to the beam collision section in electron-positron linear colliders.

Since the electron beam size with a finite beam emittance causes a rapid growth in a vacuum drift space outside plasma [41], the coupling segment must be used for spatial matching of the electron beam with the transverse wakefield as well as temporal phase matching with the accelerating wakefield in a subsequent stage. A proof-of-principle experiment on two LPA stages powered by two synchronized laser pulses through the plasma lens and mirror coupling has been reported, showing that an 120 MeV electron beam from a gas jet (the first stage) driven by a 28 TW, 45 fs laser pulse was focused by a first discharge capillary as an active plasma lens to a second capillary plasma channel (the second stage), where the wakefield excited by a 12 TW, 45 fs separate laser pulse reflected by a tape-based plasma mirror with a laser-energy throughput of 80% further increased an energy gain of 100 MeV [49]. In this experiment, a trapping fraction of the electron charge coupled to the second stage was as low as 3.5% [3]. Such a poor coupling efficiency could be attributed to the plasma mirror inserted at a vacuum drift space. To avoid a rapid growth in the vacuum drift space and improve coupling efficiency, a multistage coupling using a variable curvature plasma channel [48] enables off-axial injection of a fresh laser pulse into the LPA stage without a vacuum gap in the coupling segment; thereby an electron bunch is continuously accelerated through the plasma-focusing channel over the consecutive stages only if the temporal phase-matching between the laser and electron beams can be optimized [3].

In the propagation of a laser pulse through a curved plasma channel, the radial equilibrium position of the laser pulse is shifted away from the channel axis due to the balance between the refractive index gradient bending the light rays inward and the centrifugal force pulling them outward. As a result, the minimum of the effective plasma density, which is proportional to a guiding potential, is located outward from the channel axis [47]. Thus, a direct guiding of a laser pulse from the curved channel with a constant curvature to the straight channel causes large centroid oscillations in the straight channel even though the laser pulse is injected to the equilibrium position, leading to loss of the laser energy and electron beam transported from the previous stage as a result of off-axis interaction with plasma wakefields [48]. To diminish the mismatching at the transition from a curved channel to a straight one [3], a variable curvature plasma channel has been devised such that the equilibrium position guides the laser centroid gradually along the channel axis from the initial equilibrium position to the channel center, where the straight channel axis merges together, as shown in **Figure 7a**. A seamless acceleration in two-stage LPA coupled with a variable curvature plasma channel has been successfully demonstrated for the guided laser intensity of 8.55×10^{18} W/cm^2 (normalized vector potential $a_0 = 2$) by the three-dimensional particle-in-cell simulations, as shown in **Figure 7b–f**, indicating that the injection trapping efficiency increases with the initial beam energy and approaches 100% at energies higher than 2 GeV.

4.2 Betatron motion of the particle beam in the seamless multistage

For $s \gg 1$, the asymptotic form of betatron motion in Eq. (30) yields

$$\overline{x}(s) \sim \left[\overline{x}_0{}^2 + \left(\frac{2\gamma_0\beta_{x0}}{s_0\overline{E}_{z0}}\right)^2\right]^{1/2} \sqrt{\frac{s_0}{s}} \cos\left(s - s_0 + \delta_0\right), \tag{43}$$

$$\beta_x(s) \sim \frac{\overline{E}_z}{2\gamma}\left[\overline{x}_0{}^2 + \left(\frac{2\gamma_0\beta_{x0}}{s_0\overline{E}_{z0}}\right)^2\right]^{1/2} \sqrt{ss_0} \sin\left(s - s_0 - \delta_0\right), \tag{44}$$

where $\tan\delta_0 = 2\gamma_0\beta_{x0}/\left(s_0\overline{x}_0\overline{E}_{z0}\right)$. The variation of the betatron amplitude with respect to the initial amplitude in the k-th stage is given by

$$\left|\frac{\overline{x}(\Psi)}{\overline{x}(\Psi_{ki})}\right| \sim \sqrt{\frac{s_{ki}}{s}} = \left(\frac{\gamma_{ki}}{\gamma}\right)^{1/4} \left|\frac{\overline{E}_z(\Psi)\overline{K}(\Psi_{ki})}{\overline{E}_z(\Psi_{ki})\overline{K}(\Psi)}\right|^{1/2} \sim \left(\frac{\gamma_{ki}}{\gamma}\right)^{1/4} \left|\frac{\overline{E}_z(\Psi)}{\overline{E}_z(\Psi_{ki})}\right|^{1/2}, \tag{45}$$

where $\Psi_{ki} \le \Psi \le \Psi_{(k+1)i}$ ($k = 1, 2, \cdots$) is the particle phase Ψ with respect to the plasma wave, Ψ_{ki} is the initial phase, and γ_{ki} is corresponding to the initial energy of the particle in the k-th stage, respectively, assuming an approximately constant focusing strength $\overline{K}(\Psi) \sim \overline{K}(\Psi_{ki})$ over the stage. As expected, the betatron amplitude is simply proportional to $(\gamma_{ki}/\gamma)^{1/4}$ for the constant accelerating field $\overline{E}_z(\Psi)$ during the stage. In the two-mode mixing LPA system comprising the periodic accelerating structure, i.e., $\overline{E}_z(\Psi_{ki} + \Delta\Psi) = \overline{E}_z(\Psi_{li} + \Delta\Psi)$ for a phase advance $\Delta\Psi$ in the k-th and l-th stages, the ratio of the accelerating field amplitude is given by

$$\left|\frac{\overline{E}_z(\Psi)}{\overline{E}_z(\Psi_{ki})}\right| = e^{-2\alpha_d a_0^2(\Psi-\Psi_{ki})}\left|\frac{\Omega(\Psi)}{\Omega(\Psi_{ki})}\right|, \tag{46}$$

where $\Omega(\Psi) = \cos\Psi + \sqrt{C_1 C_2}(1+\cos 2\Psi)/(C_1+C_2)$. In the accelerator system consisting of N_s stages, the final betatron amplitude yields

$$|\overline{x}(\Psi_f)| \sim |\overline{x}_0|\left(\gamma_0/\gamma_f\right)^{1/4} R^{N_s/2}\exp\left[-\alpha_d a_0^2(\Psi_f-\Psi_0)\right], \tag{47}$$

where Ψ_f, $m_e c^2\gamma_f$ are the final phase and energy of the particle at the N_s-th stage, respectively, and $R = |\Omega(\Psi_f)/\Omega(\Psi_i)|$ is the ratio of the amplitude $\Omega(\Psi)$ between the final and initial phases in the single stage. If this ratio is chosen so as to be $R < \exp\left(2\alpha_d a_0^2\Delta\Psi\right)$, the betatron amplitude will decrease as the electron propagates the accelerator stages.

Here we consider the evolution of transverse normalized emittance for the particle beam acceleration in the multistage capillary accelerator. The definition of transverse normalized emittance given by Eq. (42) is expressed as $\varepsilon_{nx}{}^2 = \langle\delta\overline{x}^2\rangle\langle\delta u_x{}^2\rangle - \langle\delta\overline{x}\delta u_x\rangle^2$, where $\delta\overline{x} = \overline{x} - \langle\overline{x}\rangle$ and $\delta u_x = u_x - \langle u_x\rangle$ are the deviation from the mean transverse displacement $\langle\overline{x}\rangle$ and normalized momentum $\langle u_x\rangle = \langle\gamma\beta_x\rangle$, respectively. The particle orbit undergoing betatron motion is written for $s \gg 1$ from Eqs. (43) and (44) as $\overline{x} = \overline{x}_m\sqrt{s_0/s}\cos\varphi$ and $u_x = \gamma\beta_x = \overline{x}_m\left(\overline{E}_z\sqrt{ss_0}/2\right)\sin\varphi$, where

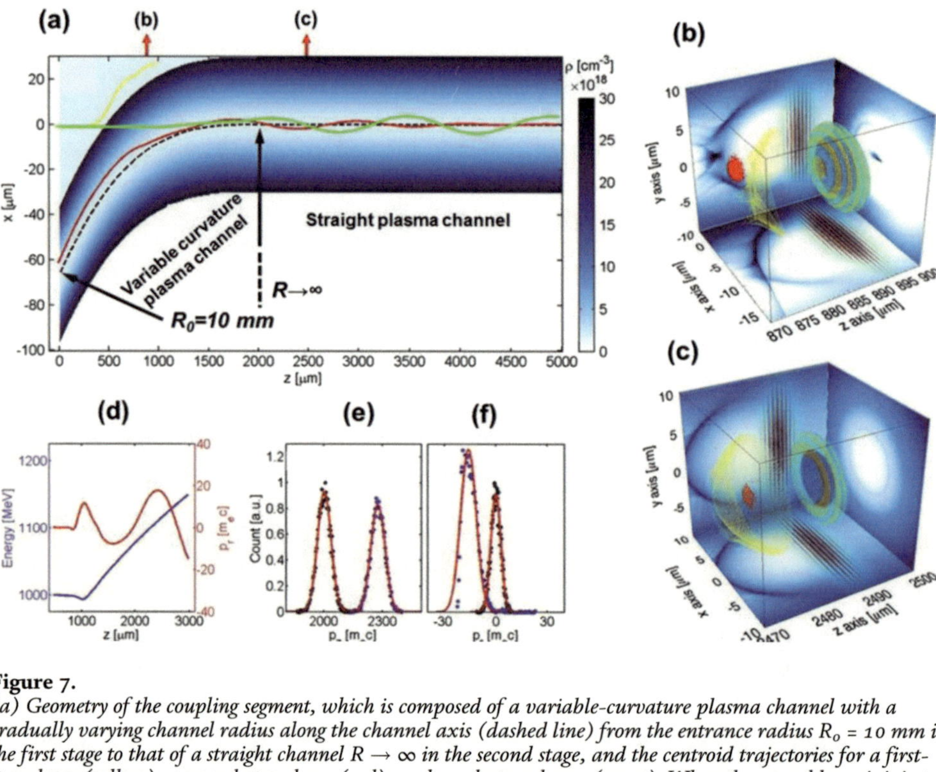

Figure 7.
(a) Geometry of the coupling segment, which is composed of a variable-curvature plasma channel with a gradually varying channel radius along the channel axis (dashed line) from the entrance radius R_0 = 10 mm in the first stage to that of a straight channel $R \to \infty$ in the second stage, and the centroid trajectories for a first-stage laser (yellow), a second-stage laser (red), and an electron beam (green). When the second laser is injected at the curved channel entrance with an incidence angle of 5.7° and an off-axis deviation of 6.3 μm, its centroid trajectory and an electron bunch (red points) are seamlessly coupled to the straight plasma channel, as shown in the three-dimensional particle-in-cell (3D PIC) simulation results before (b) and after (c) the electron bunch is trapped in the second straight channel [3]. (d) Energy and transverse momentum, initial (black points) and final (blue points) (e) longitudinal and (f) transverse momentum distributions, and their Gaussian fitting curves (red) of the electron beam obtained from the simulation results.

$\varphi = \overline{\omega}_\beta \overline{t}$ is the betatron phase and $\overline{x}_m = \left[\overline{x}_0{}^2 + \left(2\gamma_0\beta_{x0}/s_0\overline{E}_{z0}\right)^2\right]^{1/2}$. Thus, the ensemble averaged quantities $\langle \delta\overline{x}^2 \rangle$, $\langle \delta u_x{}^2 \rangle$, and $\langle \delta\overline{x}\delta u_x \rangle^2$ can be obtained: e.g., $\langle \delta\overline{x}^2 \rangle = \langle s_0/s \rangle \left[\langle \overline{x}_m{}^2 \rangle (1 + \langle \cos 2\varphi \rangle)/2 - \langle \overline{x}_m \rangle^2 \langle \cos \varphi \rangle^2\right]$. Assuming that the energy distribution about the mean energy $\langle\gamma\rangle$, i.e., the $\delta\gamma = \gamma - \langle\gamma\rangle$ distribution, is Gaussian with a width of σ_γ, the energy is approximated about its mean value to the first order in $\delta\gamma/\langle\gamma\rangle$, i.e., $\gamma = \langle\gamma\rangle + \delta\gamma$, $\omega_\beta \simeq \omega_{\beta 0}(1 - \delta\gamma/2\langle\gamma\rangle)$, and $\varphi \simeq \omega_{\beta 0}\overline{t}(1 - \delta\gamma/2\langle\gamma\rangle)$. The ensemble averaged quantities can be calculated as averages over the distribution of energy deviations as, e.g., $\langle \cos\varphi \rangle \simeq (1/\sqrt{2\pi}\sigma_\gamma)\int_{-\infty}^{\infty} d\delta\gamma \exp\left(-\delta\gamma^2/2\sigma_\gamma^2\right) \cos(\varphi_0 + \delta\varphi) = e^{-\nu_\varepsilon{}^2 t^2} \cos\varphi_0$, $\langle \cos 2\varphi \rangle \simeq e^{-4\nu_\varepsilon{}^2 t^2} \cos 2\varphi_0$, and $\langle s_0/s \rangle \simeq \left(\overline{K}_0\overline{E}_z/\overline{K}\overline{E}_{z0}\right)\left(\langle\gamma_0\rangle/\langle\gamma\rangle\right)^{1/2}$, where $\delta\varphi = -\varphi_0\delta\gamma/(2\langle\gamma\rangle)$ and $\nu_\varepsilon = \omega_{\beta 0}\sigma_\gamma/\left(\sqrt{8}\langle\gamma\rangle\right)$ is the frequency corresponding to decoherence time $t_{\text{dec}} \simeq \pi\langle\gamma\rangle/\left(\omega_\beta\sigma_\gamma\right)$, defined as the time when the phase difference between the low energy part of the beam and the high-energy part is $\Delta\varphi \simeq \omega_\beta \int dt\sigma_\gamma/\langle\gamma\rangle = \pi$ [43]. Considering transverse emittance of the particle beam with initial energy spread that dominates decoherence, the normalized emittance for $t \gg t_{\text{dec}}$ is given by

$$\overline{\varepsilon}_{nx} = \frac{1}{2}\left|\frac{\overline{E}_z}{\overline{E}_{z0}}\right| \overline{K}_0 \sqrt{\langle\gamma_0\rangle} \langle \overline{x}_{m0}{}^2 \rangle = \frac{1}{2}\left|\frac{\overline{E}_z}{\overline{E}_{z0}}\right| \overline{K}_0 \sqrt{\langle\gamma_0\rangle} \left(\langle \overline{x}_0{}^2 \rangle + \frac{\langle u_{x0}^2 \rangle}{\overline{K}_0^2 \langle\gamma_0\rangle} \right), \tag{48}$$

where $\overline{\varepsilon}_{nx} = k_p\varepsilon_{nx}$ is the dimensionless normalized emittance. If the beam is initially matched to the laser wakefield focusing channel, i.e., $\overline{x}_0 = 2\gamma_0\beta_{x0}/\left(s_0\overline{E}_{z0}\right) = u_{x0}/\left(\overline{K}_0\sqrt{\gamma_0}\right)$, such that in the absence of radiation the beam radial envelope undergoes no betatron oscillation, the normalized emittance can be expressed as

$$\overline{\varepsilon}_{nx} = \left|\frac{\overline{E}_z}{\overline{E}_{z0}}\right| \overline{K}_0 \sqrt{\langle\gamma_0\rangle} \langle x_0{}^2 \rangle = \left|\frac{\overline{E}_z}{\overline{E}_{z0}}\right| \overline{\varepsilon}_{nx0}. \tag{49}$$

This indicates that in the absence of radiation and multiple Coulomb scattering, the transverse normalized emittance of an initially matched beam is conserved in the laser wakefield acceleration when the amplitude of accelerating field has no variation, i.e., $|\overline{E}_z| = |\overline{E}_{z0}|$. Note that the decreasing accelerating field at the final phase results in a decrease of the normalized emittance of the injected beam matched to the laser wakefield at the initial phase in the single stage. For the multistage laser wakefield acceleration without a vacuum drift space in the coupling section, properly choosing the injection and extraction phases enables continuous reduction of the normalized emittance in the absence of synchrotron radiation and multiple Coulomb scattering with plasma ions. Since the initial values of the displacement $\langle\overline{x}\rangle$ and normalized momentum $\langle u_x\rangle$ at the next stage are expressed as $\langle\overline{x}_1^2\rangle \simeq \sqrt{\langle\gamma_0\rangle/\langle\gamma_1\rangle}\langle\overline{x}_m^2\rangle_0 \overline{E}_{z1}\overline{K}_0/(2\overline{E}_{z0}\overline{K}_1)$ and $\langle u_{x1}^2\rangle \simeq \overline{K}_0\overline{K}_1\sqrt{\langle\gamma_0\rangle\langle\gamma_1\rangle}\langle\overline{x}_m^2\rangle_0$ $\overline{E}_{z1}/(2\overline{E}_{z0})$ [19],

the initial amplitude of betatron oscillation at the next stage is

$$\begin{aligned}\langle\overline{x}_m^2\rangle_1 &= \langle\overline{x}_1^2\rangle + \langle u_{x1}^2\rangle/(\overline{K}_1\langle\gamma_1\rangle) \\ &= \sqrt{\langle\gamma_0\rangle/\langle\gamma_1\rangle}\langle\overline{x}_m^2\rangle_0 \overline{E}_{z1}\overline{K}_0/(\overline{E}_{z0}\overline{K}_1).\end{aligned}$$

Accordingly, the emittance at the k-th stage is calculated as

$$\bar{\varepsilon}^k{}_{nx} = (1/2)\bar{K}(\Psi_f)\sqrt{\langle\gamma\rangle_0\langle\bar{x}_m^2\rangle_0} \cdot \left|\bar{E}_z(\Psi_f)/\bar{E}_z(\Psi_i)\right|^k \left|\bar{K}(\Psi_i)/\bar{K}(\Psi_f)\right|^k.$$

Assuming $\overline{K}(\Psi_f) \approx \overline{K}(\Psi_i) = \overline{K}_0$, the dimensionless normalized emittance at the k-th stage yields

$$\bar{\varepsilon}^k{}_{nx} \simeq \frac{1}{2}\left|\frac{\overline{E}_z(\Psi_f)}{\overline{E}_z(\Psi_i)}\right|^k \overline{K}_0\sqrt{\langle\gamma\rangle_0\langle\overline{x}_m^2\rangle_0} = \frac{1}{2}R^k e^{-2\alpha a_0^2 k\Delta\Psi}\overline{K}_0\sqrt{\langle\gamma\rangle_0\langle\overline{x}_m^2\rangle_0} = R^k e^{-2\alpha a_0^2 k\Delta\Psi}\overline{\varepsilon}_{n0}, \tag{50}$$

where $\left|\overline{E}_z(\Psi_f)/\overline{E}_z(\Psi_i)\right| = Re^{-2\alpha a_0^2\Delta\Psi}$ is the ratio of the accelerating field amplitude at the final phase Ψ_f to that at the initial phase Ψ_i with $R = \left|\Omega(\Psi_f)/\Omega(\Psi_i)\right|$ and $\Delta\Psi = \Psi_f - \Psi_i$. Setting $\overline{K}_0\langle\gamma_0\rangle^{1/2}\langle\overline{x}_m^2\rangle = 2\overline{\varepsilon}_{n0}$, the normalized emittance increases or decreases monotonically, depending on $R > e^{2\alpha a_0^2\Delta\Psi}$ or $R < e^{2\alpha a_0^2\Delta\Psi}$ as the particles move along the stage in the absence of radiation and multiple Coulomb scattering.

For an application of laser-plasma accelerators to electron-positron colliders, it is of most importance to achieve the smallest possible normalized emittance at the final stage of the accelerator system, overwhelming the emittance growth due to the multiple Coulomb scattering off plasma ions, being increased in proportion to the square root of the beam energy. We consider the effect of multiple Coulomb scattering on the emittance growth and evaluate an achievable normalized emittance at the end of the accelerator system comprising N_s stages. Using the growth rate of the mean square deflection angle in Eq. (36) due to the multiple Coulomb scattering, the growth rate of the transverse normalized emittance is estimated as [8, 44].

$$\frac{d\varepsilon_n^{\rm SCAT}}{dz} = \frac{1}{2}\frac{\gamma}{k_\beta}\frac{d\langle\theta^2\rangle}{dz} = \frac{k_p r_e Z}{\overline{K}\sqrt{\gamma}}\ln\left(\frac{\lambda_D}{R_N}\right), \tag{51}$$

where $k_\beta = K/\sqrt{\gamma}$ is the wave number of betatron oscillation. In the single stage, the transverse normalized emittance of the particles undergoing the wakefields evolves the growth in the same manner as the injected particle beam without the multiple Coulomb scattering as

$$\bar{\varepsilon}^1{}_{nx} = \left|\frac{\overline{E}_z(\Psi_1)}{\overline{E}_z(\Psi_0)}\right|\overline{\varepsilon}_{n0} + \frac{k_p r_e Z\ln(\lambda_D/R_N)}{\overline{K}_0\overline{E}_z(\Psi_0)}\left(\sqrt{\langle\gamma_{1f}\rangle} - \sqrt{\langle\gamma_{1i}\rangle}\right). \tag{52}$$

At the N_s-th stage, the normalized emittance can be obtained from

$$\bar{\varepsilon}^N{}_{nx} = \left|\frac{\overline{E}_z(\Psi_1)}{\overline{E}_z(\Psi_0)}\right|^N \overline{\varepsilon}_{n0} + \frac{k_p r_e Z\ln(\lambda_D/R_N)}{\overline{K}_0\overline{E}_z(\Psi_0)}\sum_{k=1}^{N}\left|\frac{\overline{E}_z(\Psi_1)}{\overline{E}_z(\Psi_0)}\right|^{N-k}\sqrt{\gamma_k}\left(1-\sqrt{\frac{\gamma_{k-1}}{\gamma_k}}\right). \tag{53}$$

Assuming that the beam energy at the k-th stage is approximately given by $\langle\gamma_k\rangle \simeq (\pi C_1 C_2/2)^{1/2} a_0^2\gamma_g^2 e^{-(1+4\alpha a_0^2\Psi_0)/2} k\Delta\Psi$ for $k \gg 1$, Eq. (53) can be calculated as

$$\begin{aligned}\bar{\varepsilon}^N{}_{nx} = {}& \overline{\varepsilon}_{n0}R^{N_s}e^{-2N_s\alpha_d a_0^2\Delta\Psi} + \left(\frac{C_1C_2}{2\pi}\right)^{1/4}\frac{C_S Z\ln(\lambda_D/R_N)}{a_0(C_1+C_2)}\frac{e^{(1+4\alpha a_0^2\Psi_0)/4}R^{N_s}e^{-2N_s\alpha_d a_0^2\Delta\Psi}\sqrt{\Delta\Psi}}{\overline{K}_0|\Omega(\Psi_0)|\sqrt{\ln R - 2\alpha a_0^2\Delta\Psi}} \\ & \times\left[\operatorname{erf}\sqrt{N_s(\ln R - 2\alpha_d a_0^2\Delta\Psi)} - \operatorname{erf}\sqrt{\ln R - 2\alpha_d a_0^2\Delta\Psi}\right],\end{aligned} \tag{54}$$

for $R > e^{2\alpha a_0^2 \Delta\Psi}$, where $C_S = k_p r_e \gamma_g = 17.7 \times 10^{-9} [\mu\text{m}]/\lambda_0$ and $\text{erf}(x) = (2/\sqrt{\pi}) \int_0^x e^{-t^2} dt$ are the error function, and for $R < e^{2\alpha a_0^2 \Delta\Psi}$,

$$\overline{\varepsilon}^N{}_{nx} = \overline{\varepsilon}_{n0} R^{N_s} e^{-2N_s \alpha a_0^2 \Delta\Psi} + \left(\frac{2}{\pi}\right)^{3/4} \frac{C_s (C_1 C_2)^{1/4} Z \ln(\lambda_D / R_N)}{a_0 (C_1 + C_2)} \frac{e^{(1+4\alpha_d a_0^2 \Psi_0)/4} \sqrt{\Delta\Psi}}{\overline{K}_0 |\Omega(\Psi_0)| \sqrt{2\alpha_d a_0^2 \Delta\Psi - \ln R}},$$
$$\times \left\{ \text{F}\left[\sqrt{N_s (2\alpha_d a_0^2 \Delta\Psi - \ln R)}\right] - R^{N_s - 1} e^{-2(N_s - 1)\alpha_d a_0^2 \Delta\Psi} \text{F}\left[\sqrt{2\alpha_d a_0^2 \Delta\Psi - \ln R}\right] \right\} \tag{55}$$

where $\text{F}(x) = e^{-x^2} \int_0^x e^{t^2} dt$ is Dawson's integral. For $R = e^{2\alpha a_0^2 \Delta\Psi}$, i.e., $|\overline{E}_z(\Psi_f)/\overline{E}_z(\Psi_i)| = 1$, the normalized emittance at the N_s-th stage is simply calculated as

$$\begin{aligned} \overline{\varepsilon}^N{}_{nx} &= \overline{\varepsilon}_n{}^0 + \frac{k_p r_e Z \ln(\lambda_D / R_N)}{\overline{K}_0 |\overline{E}_z(\Psi_0)|} \left(\sqrt{\gamma_{N_s}} - \sqrt{\gamma_0}\right) \\ &\sim \overline{\varepsilon}_n{}^0 + 2\left(\frac{2C_1 C_2}{\pi}\right)^{1/4} \frac{C_S \beta_g Z \ln(\lambda_D / R_N)}{a_0 (C_1 + C_2)} \frac{e^{(1+4\alpha a_0^2 \Psi_0)/4} \sqrt{N_s \Delta\Psi}}{\overline{K}_0 |\Omega(\Psi_0)|}. \end{aligned} \tag{56}$$

As expected, the normalized emittance in the multistage accelerator operated with the constant accelerating field is conserved to the initial normalized emittance and then limited by an increasing growth due to multiple Coulomb scattering. For $R > e^{2\alpha a_0^2 \Delta\Psi}$, the initial emittance of the injected beam dominates an exponential growth of the normalized emittance, while for $R < e^{2\alpha a_0^2 \Delta\Psi}$, an exponential decrease of the initial emittance is followed by a slow decrease of the normalized emittance arising from the multiple Coulomb scattering [19].

4.3 Numerical studies of the single-particle dynamics in multistages

Numerical studies on transverse beam dynamics of an electron bunch accelerated in the multistage mode mixing LPA have been carried out by calculating the ensemble of trajectories of test particles throughout consecutive stages, using the single-particle dynamics code based on the analytical solutions of the equations of motion of an electron in laser wakefields with the presence of effects of the radiation reaction and multiple Coulomb scattering, as described in Section 3. **Figure 8a** shows examples of the phase space distribution of 10^4 test particles on the $k_p x - \gamma\beta_x$ plane and evolution of the transverse normalized emittance for 400 stages, in each of which the electron is accelerated between the initial wakefield phase $\Psi_i = 0$ and final phase $\Psi_f = 4.5\pi$, in the presence of the radiation reaction and multiple Coulomb scattering. **Figure 8b** is the result for 60 stages with the stage phase $0 \le \Psi \le 4\pi$ and **Figure 8c** for 50 stages with $-0.45\pi \le \Psi \le 4\pi$, taking into account only the radiation effect. The cases shown in **Figure 8a** and **b** obviously correspond to the exponential decrease of the normalized emittance with $R < \exp\left(2\alpha a_0^2 \Delta\Psi\right)$, while the case shown in **Figure 8c** corresponds to the exponential increase with $R > \exp\left(2\alpha a_0^2 \Delta\Psi\right)$. In **Figure 8a**, the exponential decrease of the normalized emittance is limited, leading to the equilibrium with the growth due to the multiple Coulomb scattering after several stages. In **Figure 8c**, the exponential increase can be limited by the radiation effects, resulting in an excess of radiation energy loss and the equilibrium with the radiation reaction after 20 stages. For the case shown in **Figure 8a**, the beam energy is accelerated up to 558.92 GeV with the relative rms

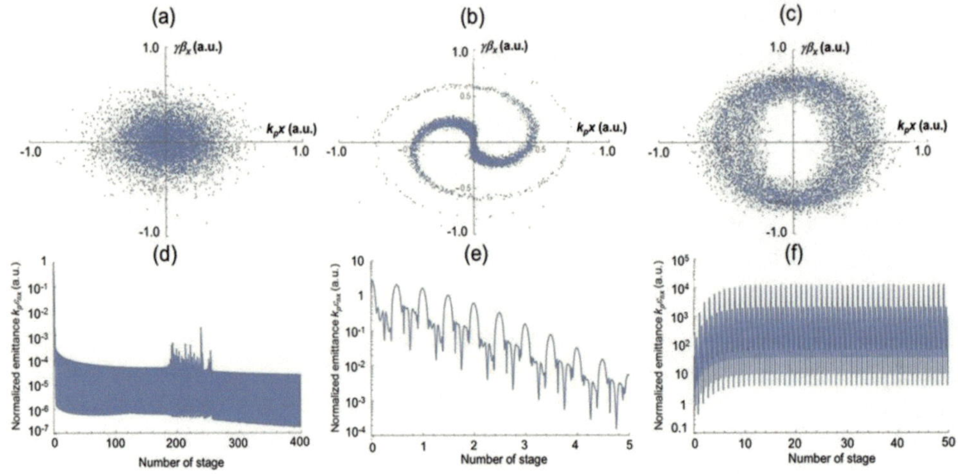

Figure 8.
*Numerical results of the beam dynamics study on the two-mode mixing multistage LPA (**Figure 1a**). (a and d) the phase space $k_p x - \gamma\beta_x$ and evolution of transverse normalized emittance $k_p\varepsilon_{nx}$ for 400 stages with each stage phase $0 \leq \Psi \leq 4.5\pi$ in the presence of radiation reaction and multiple coulomb scattering at the plasma density $n_e = 1 \times 10^{18}$ cm^{-3} and the initial normalized emittance $\varepsilon_{no} = 1$ μm. (b and e) the phase space and evolution of transverse normalized emittance (only shown for the initial 5 stages) for 60 stages with each stage phase $0 \leq \Psi \leq 4\pi$ in the presence of the radiation reaction at $n_e = 1.1 \times 10^{17}$ cm^{-3} and $\varepsilon_{no} = 100$ μm. (c and f) the phase space and evolution of transverse normalized emittance for 50 stages with each stage phase $-0.45\pi \leq \Psi \leq 4\pi$ in the presence of the radiation reaction at $n_e = 1.1 \times 10^{17}$ cm^{-3} and $\varepsilon_{no} = 100$ μm.*

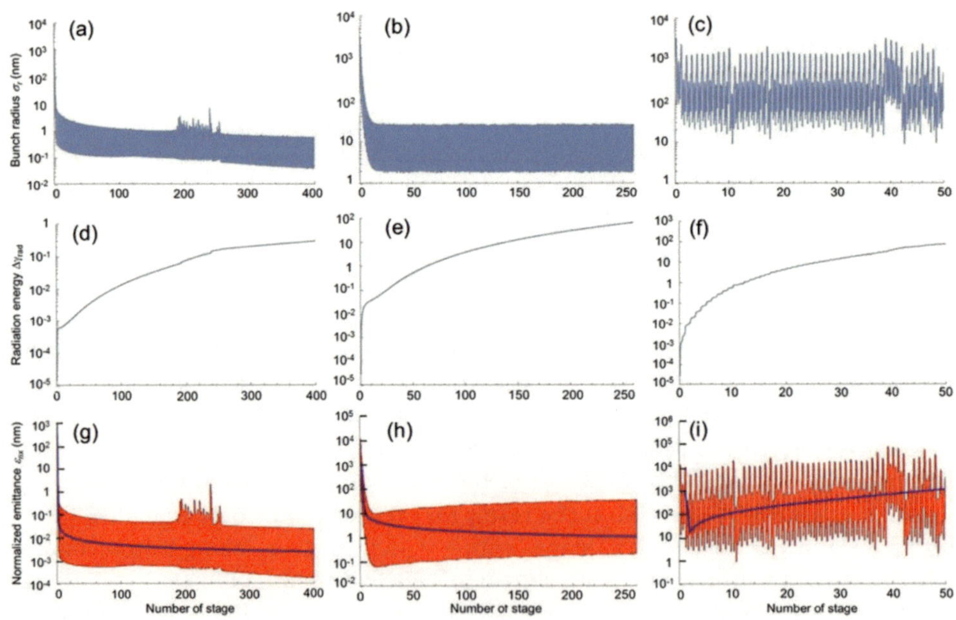

Figure 9.
Numerical results of the beam dynamics study on the two-mode mixing multistage LPA at the plasma density $n_e = 1 \times 10^{18}$ cm^{-3} and number of electrons $N_b = 1 \times 10^8$ in a bunch with a length of $k_p\sigma_z = 3$, the initial energy $m_e c^2\gamma_0 = 1$ GeV, relative energy spread $\sigma_\gamma/\gamma_0 = 0.1$, and normalized emittance $\varepsilon_{no} = 1$ μm. (a, d, and g) evolution of rms bunch radius σ_r, radiation energy $\Delta\gamma_{rad}$, and transverse normalized emittance ε_{nx} for 400 stages with each stage phase $0 \leq \Psi \leq 4.5\pi$. (b, e, and h) evolution of the same electron beam parameters for 260 stages with each stage phase $-0.45\pi \leq \Psi \leq 4.7\pi$. (c f, and i) evolution of the same electron beam parameters for 50 stages with each stage phase $-0.45\pi \leq \Psi \leq 4.2345\pi$. In (g), (h), and (i), the blue solid curve shows a fit using the analytical emittance evolution formula in Eqs. (54) and (55).

energy spread of 0.02% over the whole 400 LPA stages with the stage phase $0 \leq \Psi \leq 4.5\pi$ at the operating plasma density of $n_e = 1 \times 10^{18}$ cm^{-3} in the accelerator length of 67 m, assuming initially a 10% relative energy spread. From this result, the beam-induced longitudinal (decelerating) wakefield becomes approximately $E_{zb}/E_0 \sim 0.01$ and focusing strength $(K_b/k_p)^2 \sim 0.12$, as calculated from an ensemble average of the radius of an electron bunch with 1×10^8 electrons and length of 16 μm, using Eqs. (20) and (26), respectively, at each step of the stage consisting of 100 segments. It is noted that both beam-induced wakefields reach the equilibrium after several stages in consistency with the evolution of the normalized emittance.

The detailed study on the evolution of the transverse normalized emittance in the multistage two-mode mixing LPA has been investigated for three cases with the different stage phases, i.e., $0 \leq \Psi \leq 4.5\pi$ (case A), $-0.45\pi \leq \Psi \leq 4.7\pi$ (case B), and $-0.45\pi \leq \Psi \leq 4.2345\pi$ (case C), the reduction coefficients $2\alpha_d a_0^2 \Delta\Psi - \ln R$ for which are 37.3, 1.37, and $-$ 0.438, respectively. **Figure 9** shows the evolution of the bunch radius σ_r (a)–(c), radiation energy $\Delta\gamma_{\rm rad}$ (d)–(f), and transverse normalized emittance ε_{nx} (g)–(i), respectively. In **Figure 9** (g)–(i), the solid curve indicates the normalized emittance predicted from the analytical formulae of Eqs. (54) and (55), assuming that the average focusing constant is $\overline{K}_0 \sim 0.004$ for cases A–C and the growth rate is ~10% of the reduction coefficient for case C. It is noted that the evolution of the normalized emittance is determined by the equilibrium between the consecutive focusing and the defocusing due to the multiple Coulomb scattering at a large number of stages $N_s \gg 1$.

5. Considerations on electron-positron collider performance

Electron and positron beams being reached to the final energies in the multistage two-mode mixing LPA are extracted at a phase corresponding to the minimum transverse normalized emittance, followed by propagating a drift space in vacuum and a final focusing system to the beam-beam collisions at the interaction point. In a vacuum drift space outside plasma, the particle beam changes the spatial and temporal dimensions of the bunch proportional to the propagation distance due to the finite emittance and energy spread of the accelerated bunch. The evolution of the rms bunch envelope σ_b in vacuum without the external focusing force is given by $\sigma_b^2 = \sigma_0^2\left[1 + (z - z_0)^2/Z_b^2\right]$, where σ_0 is an initial radius and $Z_b = \sigma_0^2\gamma/\varepsilon_n$ is the characteristic distance of the bunch size growth [41]. The bunch radius after propagation of the distance $L_{\rm col}$ between the final LPA stage and interaction point is estimated to be $\sigma_{b^*} = \left[\sigma_{bf}^2 + \left(\varepsilon_{nf}L_{\rm col}/\gamma\sigma_{bf}\right)^2\right]^{1/2}$, where σ_{b^*} is the rms bunch radius at the interaction point and σ_{bf}, ε_{nf} the rms bunch radius and normalized emittance at the exit of the LPA, respectively. In collisions between high-energy electron and positron bunches from the LPAs, the beam particles emit synchrotron radiation due to the interaction, the so-called Beamstrahlung, with the electromagnetic fields generated by the counterpropagating beam. The beamstrahlung effect leads to substantial beam energy loss and degradation on energy resolution for the high-energy experiments in electron-positron linear colliders [50]. Intensive research on beamstrahlung radiation has been explored [50–52], being of relevance to the design of e^+e^- linear colliders in the TeV center-of-mass (CM) energies, for which two major effects must be taken into account, namely, the disruption effect bending particle trajectories by the oncoming beam-generated electromagnetic fields and the beamstrahlung effect yielding radiation loss of the particle energies induced by bending their trajectories due to the disruption [52]. The radiative energy loss due to beamstrahlung for a Gaussian beam can be estimated in terms of the beamstrahlung parameter $\Upsilon_* = 5r_e{}^2\gamma N_b/(12\alpha\sigma_x^*\sigma_z^*)$ for a round beam with $\sigma_{x*} = \sigma_{y*}$, where $\alpha = e^2/\hbar c$ ($\simeq 1/137.036$) is the fine-structure constant [50]. According to the beamstrahlung simulations [50], the average number of emitted photon per electron and average fractional energy loss are $n_\gamma \approx 2.54 B_\gamma \Upsilon_*/(1+\Upsilon_*^{2/3})^{1/3}$ and $\delta_b \approx 1.24 B_\gamma \Upsilon_*^2/\left[1+(3\Upsilon_*/2)^{2/3}\right]^2$, respectively, with $B_\gamma = \alpha^2\sigma_{z^*}/(r_e\gamma)$. Using these parameters, the average CM energy loss can be calculated as $\Delta W/(2\gamma m_e c^2) \approx (0.44 + 0.01\log_{10}\Upsilon_*)\delta_b(1+\delta_b/n_\gamma)$. In the quantum beamstrahlung regime, the collider design must consider the CM energy loss such that their requirements can be reached as well as that of the luminosity. The geometric luminosity is given by $\mathcal{L}_0 = f_c N_b^2/(4\pi\sigma_{x^*}\sigma_{y^*})$, where f_c is the collision frequency. It is pointed out that an appreciable disruption effect turns out to the luminosity enhancement through the pinch effect arising from the attraction of the oppositely charged beams [51, 52]. For Gaussian beams, the disruption parameter for the round beam is $D = r_e\sigma_{z^*}N_b/(\gamma\sigma_{x^*}^2)$, defined as the ratio of the bunch length to the focal length of a thin lens. The luminosity enhancement factor being defined as the ratio of the effective luminosity $\mathcal{L}$ induced by the disruption to the geometric luminosity in the absence of disruption $\mathcal{L}_0$ is estimated from the empirical formula: $H_D \equiv \mathcal{L}/\mathcal{L}_0 = 1 + D^{1/4}(D^3/1+D^3)\left[\ln(\sqrt{D}+1) + 2\ln(0.8/A)\right]$, where $A = \varepsilon_n D/(r_e N)$ is the inherent divergence of the incoming beam [52]. This scenario allows us to transport both

Plasma density n_e	1×10^8 cm^{-3}
Plasma wavelength λ_p	33.4 μm
Capillary radius R_c	152.6 μm
Capillary stage length	16.75 cm
Laser wavelength λ	1 μm
Laser spot radius r_0	91 μm (51 μm)
Laser pulse duration τ	25 fs
Normalized vector potential a_0	1
Electromagnetic hybrid mode	EH_{11} and EH_{12}
Coupling efficiency for an Airy beam (a Gaussian beam)	C_1 = 0.4022 (0.5980) C_2 = 0.4986 (0.3531)
Bunch initial and final phase	$\Psi_i = 0$, $\Psi_f = 4.5\pi$
Average accelerating gradient	8.3 GeV/m
Laser peak power P_L	95 TW (18 TW)
Laser pulse energy U_L	2.4 J (0.44 J)
Repetition frequency f_c	50 kHz
Laser average power per stage	120 kW (22 kW)
Laser depletion η_{pd}	77%

The parameters in () correspond to the incident laser pulse with a Gaussian beam.

Table 1.
Parameters of the two-mode mixing laser-plasma accelerator stage.

CM energy	1.12 TeV
Beam energy	559 GeV
Injection beam energy	1 GeV
Particle per bunch N_b	1×10^8
Collision frequency f_c	50 kHz
Total beam power	0.9 MW
Geometric luminosity $\mathfrak{L}_0$	3.6×10^{32} cm^{-2} s^{-1}
Effective luminosity $\mathfrak{L}$	1.76×10^{34} cm^{-2} s^{-1}
Effective CM energy	1.09 TeV
rms CM energy spread	8.4%
rms bunch length σ_z	16 μm
Beam radius at IP σ_{b^*}	3.3 nm
Beam aspect ratio R	1
Normalized emittance at IP ε_{nf}	3.7 pm-rad
Distance between LPA and IP L_{col}	0.2 m
Beamstrahlung parameter Υ_*	0.94
Beamstrahlung photons n_γ	0.52
Disruption parameter D	12
Luminosity enhancement H_D	49
Number of stages per beam N_s	400
Linac length per beam	67 m
Power requirement for lasers	95 MW (18 MW)

The parameters in () correspond to the incident laser pulse with a Gaussian beam.

Table 2.
Parameters for 1 TeV laser-plasma e^+e^- linear collider.

beams into the interaction point through no extra focusing devices, which often induce the degradation of beam qualities prior to their interactions. In this scheme, the vacuum drift region from the end of the LPA to the interaction point can be used for control of the transverse beam size that strongly affect the luminosity and CM energy through the beamstrahlung radiation and disruption. A typical design example of the LPA stage using the gas-filled [19] two-mode mixing LPA operated with EH_{11} and EH_{12} is shown in **Table 1**.

An embodiment of the LPA stage may be envisioned by exploiting a tens kW-level high-average power laser such as a coherent amplification network of fiber lasers [53]. **Table 2** summarizes key parameters on the performance of 1 TeV CM energy electron-positron linear collider.

6. Conclusions

The electron acceleration and beam dynamics of the two-mode mixing LPA comprising a gas-filled metallic or dielectric capillary have been presented for the performance of the single-stage and multistage configurations. As shown in **Table 1**, when a laser pulse with an Airy beam (or a Gaussian beam) profile of the spot radius $r_0 \simeq 0.6R_c$ ($r_0 = 0.33R_c$) is coupled to a gas-filled capillary, two electromagnetic hybrid modes EH_{11} and EH_{12} are generated with the coupling efficiency $C_1 = 0.40$ ($C_1 = 0.60$) and $C_2 = 0.50$ ($C_2 = 0.35$), respectively. Furthermore, when the capillary radius is tuned to the matching condition given by Eq. (12), the laser pulse comprising two beating hybrid modes EH_{11} and EH_{12} with a Gaussian temporal profile can efficiently excite a rectified accelerating wakefield, where relativistic electrons dominantly propagate in the accelerating phase and continuously gain the energy until depletion of the laser pulse energy, whereby a nearly 100% of the laser energy can be transferred to wakefields in the single stage.

In the two-mode mixing LPA multistage coupled with a variable curvature plasma channel, the transverse dynamics of the electron bunch is dominated by seamless recurrence of the accelerating wakefield in the stages, where the cumulative nature of the particle trajectories is determined by the amplitude ratio of the accelerating field at the final phase Ψ_f to the initial phase Ψ_i in each stage, i.e., $|\overline{E}_z(\Psi_f)/\overline{E}_z(\Psi_i)|$. With the converging condition, i.e., $|\overline{E}_z(\Psi_f)/\overline{E}_z(\Psi_i)| < 1$, the bunch radius and normalized emittance exhibit an exponential decrease initially and then turn out to be in equilibrium with the growth due to the multiple Coulomb scattering after 20 stages, leading to the rms bunch radius of the order of ~1 nm and the transverse normalized emittance of the order of ~0.1 nm-rad at the beam energy 559 GeV with the relative rms energy spread of 0.02% in the final 400 stage of the accelerator length of 67 m, as shown for case A in **Figure 9**. This capability of producing such high-energy and high-quality electron (or positron) beams allows us to conceive a unique electron-positron linear collider with high luminosity of the order of 10^{34} cm^{-2} s^{-1} at 1 TeV center-of-mass energy in a very compact size.

In conclusion, a novel scheme of 1 TeV electron-positron linear collider comprising properly phased multistage two-mode mixing LPAs using gas-filled capillary waveguides can provide a unique approach in collider applications. This scheme presented resorts two major mechanisms pertaining to laser wakefield acceleration, that is, dephasing and strong focusing force as well as very high-gradient accelerating field. The multistage scheme using two-mode mixing capillary waveguides filled with plasma may provide a robust approach leading to the supreme goal for LPAs. The numerical model developed for study on beam dynamics in large-scale LPAs will be useful for assessing effects of underlying physics and the optimum design for future laser-plasma-based colliders. Although the present model has been developed to study the simplest two-dimensional phase-space model of electron beam dynamics in laser wakefield acceleration, the analysis of higher multi-dimensional phase-space model as well as the quantum plasma effect will be extensively pursued in the future work.

Acknowledgements

The work was supported by the NSFC (No. 11721091, 11774227), the Science Challenge Project (No.TZ2018005), and the National Basic Research Program of China (No. 2013CBA01504).

Author details

Kazuhisa Nakajima[1*], Min Chen[1] and Zhengming Sheng[1,2]

1 Key Laboratory for Laser Plasmas (MOE), School of Physics and Astronomy, Shanghai Jiao Tong University, Shanghai, China

2 SUPA, Department of Physics, University of Strathclyde, Glasgow, UK

*Address all correspondence to: naka115@dia-net.ne.jp

References

[1] Gibney E. Who will build the next LHC? Nature. 2016;**536**:383-384

[2] Schulte D. Application of advanced accelerator concepts for colliders. In: Chao AW, Chou W, editors. Review of Accelerator Science and Technology. Vol. 9. Singapore: World Scientific; 2016. pp. 209-233. DOI: 10.1142/S1793626816300103

[3] Nakajima K. Seamless multistage laser-plasma acceleration toward future high-energy colliders. Light: Science & Applications. 2018;**7**:21. DOI: 10.1038/s41377-018-0037-6

[4] Tajima T, Dawson JM. Laser electron accelerator. Physical Review Letters. 1979;**43**:267-270. DOI: 10.1103/PhysRevLett.43.267

[5] Nakajima K. In: Chao AW, Chou W, editors. Review of Accelerator Science and Technology. Vol. 9. Singapore: World Scientific; 2016. pp. 19-61. DOI: 10.11412/S1793626816300024

[6] Grüner F, Becker S, Schramm U, Eichner T, Fuchs M, Weingartner R, et al. Design considerations for table-top, laser-based VUV and X-ray free electron lasers. Applied Physics B: Lasers and Optics. 2007;**86**:431-435. DOI: 10.1007/s00340-006-2565-7

[7] Nakajima K. Compact X-ray sources-towards a table-top free-electron laser. Nature Physics. 2008;**4**:92-93

[8] Schroeder CB, Esarey E, Geddes CGR, Benedetti C, Leemans WP. Physics considerations for laser-plasma linear colliders. Physical Review Accelerators and Beams. 2010; **13**:101301. DOI: 10.1103/PhysRevSTAB.13.101301

[9] Nakajima K, Deng A, Zhang X, Shen B, Liu J, Li R, et al. Operating plasma density issues on large-scale laser-plasma accelerators toward high-energy frontier. Physical Review Accelerators and Beams. 2011;**14**:091301. DOI: 10.1103/PhysRevSTAB.14.091301

[10] Nakajima K, Lu HY, Zhao XY, Shen BF, Li RX, Xu ZZ. 100-GeV large scale laser plasma electron acceleration by a multi-PW laser. Chinese Optics Letters. 2013;**11**:013501. DOI: 10.3788/COL201311.013501

[11] Liu JS, Xia CQ, Wang WT, Lu HY, Wang C, Deng AH, et al. All-optical cascaded laser wakefield accelerator using ionization-induced injection. Physical Review Letters. 2011;**107**:035001. DOI: 10.1103/PhysRevLett.107.035001

[12] Wang X, Zgadzaj R, Fazel N, Li Z, Yi SA, Zhang X, et al. Quasi-monoenergetic laser-plasma acceleration of electrons to 2 GeV. Nature Communications. 2013;**4**:1988-1-1988-9. DOI: 10.1038/ncomms2988

[13] Kim HT, Pae KH, Cha HJ, Kim IJ, Yu TJ, Sung JH, et al. Enhancement of electron energy to the multi-GeV regime by a dual-stage laser-wakefield accelerator pumped by petawatt laser pulses. Physical Review Letters. 2013; **111**:165002. DOI: 10.1103/PhysRevLett.111.165002

[14] Kim HT, Pathak VB, Pae KH, Lifschitz A, Sylla F, Shin JH, et al. Stable multi-GeV electron accelerator driven by waveform-controlled PW laser pulses. Scientific Reports. 2017;**7**:1020. DOI: 10.1038/s41598-017-09267-1

[15] Leemans WP, Gonsalves AJ, Mao HS, Nakamura K, Benedetti C, Schroeder CB, et al. Multi-GeV electron beams from capillary-discharge-guided subpetawatt laser pulses in the self-trapping regime. Physical Review Letters. 2014;**113**:245002. DOI: 10.1103/PhysRevLett.113.245002

[16] Lu H, Liu M, Wang W, Wang C, Liu J, Deng A, et al. Laser wakefield acceleration of electron beams beyond 1 GeV from an ablative capillary discharge waveguide. Applied Physics Letters. 2011;**99**:091502. DOI: 10.1063/1.3626042

[17] Gonsalves AJ, Nakamura K, Daniels J, Benedetti C, Pieronek C, de Raadt TCH, et al. Petawatt laser guiding and electron beam acceleration to 8 GeV in a laser-heated capillary discharge waveguide. Physical Review Letters. 2019;**122**:084801. DOI: 10.1103/PhysRevLett.122.084801

[18] Osterhoff J, Popp A, Major Z, Marx B, Rowlands-Rees TP, Fuchs M, et al. Generation of stable low-divergence electron beams by laser-wakefield acceleration in a steady-state-flow gas cell. Physical Review Letters. 2008;**101**:085002. DOI: 10.1103/PhysRevLett.101.085002

[19] Nakajima K, Wheeler J, Mourou G, Tajima T. Novel laser-plasma TeV electron-positron linear colliders. International Journal of Modern Physics A. 2019;**34**:1943003. DOI: 10.1142/S0217751X19430036

[20] Cros B, Courtois C, Matthieussent G, Bernardo AD, Batani D, Andreev N, et al. Eigenmodes for capillary tubes with dielectric walls and ultraintense laser pulse guiding. Physical Review E. 2002;**65**:026405. DOI: 10.1103/PhysRevE.65.026405

[21] Nakajima K. Laser electron acceleration beyond 100 GeV. The European Physical Journal Special Topics. 2014;**223**:999-1016. DOI: 10.1140/epjst/e2014-02151-9

[22] Behnke T, Brau TJE, Foster B, Fuster J, Harrison M, Paterson JM, et al., editors. The international linear collider accelerator. In: The International Linear Collider Technical Design Report. Vol. 1. 2013. pp. 9-26

[23] Durfee CG III, Milchberg HM. Light pipe for high intensity laser pulses. Physical Review Letters. 1993;**71**:2409. DOI: 10.1103/PhysRevLett.71.2409

[24] Xiao YF, Chu HH, Tsai HE, Lee CH, Lin JY, Wang J, et al. Efficient generation of extended plasma waveguides with the axicon ignitor-heater scheme. Physics of Plasmas. 2004;**11**:L21-L24. DOI: 10.1063/1.1695354

[25] Levin M, Pukhov A, Zigler A, Sugiyama K, Nakajima K, Hubbard RF, et al. Long plasma channels in segmented capillary discharges. Physics of Plasmas. 2006;**13**:083108. DOI: 10.1063/1.2261853

[26] Liu MW, Deng AH, Liu JS, Li RX, Xu JC, Xia CQ, et al. Note: Low density and long plasma channels generated by laser transversely ignited ablative capillary discharges. The Review of Scientific Instruments. 2009;**81**:036107. DOI: 10.1063/1.3360925

[27] Hosokai T, Kando M, Dewa H, Kotaki H, Kondo S, Hasegawa N, et al. Optical guidance of terawatt laser pulses by the implosion phase of a fast Z-pinch discharge in a gas-filled capillary. Optics Letters. 2000;**25**:10-12

[28] Gonsalves AJ, Rowlands-Rees TP, Broks BHP, van der Mullen JJAM, Hooker SM. Transverse interferometry of a hydrogen-filled capillary discharge waveguide. Physical Review Letters. 2007;**98**:025002. DOI: 10.1103/PhysRevLett.98.025002

[29] Gonsalves AJ, Liu F, Bobrova NA, Sasorov PV, Pieronek C, Daniels J, et al. Demonstration of a high repetition rate capillary discharge waveguide. Journal of Applied Physics. 2016;**119**:033302. DOI: 10.1063/1.4940121

[30] Ju J, Svensson K, Döpp A, Ferrari HE, Cassou K, Neveu O, et al. Enhancement of x-rays generated by a

guided laser wakefield accelerator inside capillary tubes. Applied Physics Letters. 2012;**100**:191106. DOI: 10.1063/1.4712594

[31] Curcio A, Petrarca M, Giulietti D, Ferrario M. Numerical and analytical models to study the laser-driven plasma perturbation in a dielectric gas-filled capillary waveguide. Optics Letters. 2016;**41**:4233-4236. DOI: 10.1364/OL.41.004233

[32] Du D, Liu X, Korn G, Squier J, Mourou G. Laserinduced breakdown by impact ionization in SiO2 with pulse widths from 7 ns to 150 fs. Applied Physics Letters. 1994;**64**:3071-3073. DOI: 10.1063/1.111350

[33] Veysman M, Andreev NE, Cassou K, Ayoul Y, Maynard G, Cros B. Theoretical and experimental study of laser beam propagation in capillary tubes for non-symmetrical coupling conditions. Journal of the Optical Society of America B: Optical Physics. 2010;**27**:1400-1408

[34] Andreev NE, Cros B, Maynard G, Mora P, Wojda F. Coupling efficiency of intense laser pulses to capillary tubes for laser wakefield acceleration. IEEE Transactions on Plasma Science. 2008; **36**:1746-1750. DOI: 10.1109/TPS.2008.927144

[35] Esarey E, Schroeder CB, Leemans WP. Physics of laser-driven plasma-based electron accelerators. Reviews of Modern Physics. 2009;**81**: 1229-1284. DOI: 10.1103/RevModPhys.81.1229

[36] Sprangle P, Esarey E, Ting A. Nonlinear theory of intense laser-plasma interactions. Physical Review Letters. 1990;**64**:2011-2014

[37] Shadwick BA, Schroeder CB, Esarey E. Nonlinear laser energy depletion in laser-plasma accelerators. Physics of Plasmas. 2009;**16**:056704. DOI: 10.1063/1.3124185

[38] Lu W, Huang C, Zhou MM, Mori WB, Katsouleas T. Limits of linear plasma wakefield theory for electron or positron beams. Physics of Plasmas. 2005;**12**:063101. DOI: 10.1063/1.1905587

[39] Panofsky WKH, Wenzel WA. Some considerations concerning the transverse deflection of charged particles in radio-frequency fields. The Review of Scientific Instruments. 1956; **27**:967. DOI: 10.1063/1.1715427

[40] Chen P, Su JJ, Katsouleas T, Wilks S, Dawson JM. Plasma focusing for high-energy beams. IEEE Transactions on Plasma Science. 1987; **PS-15**:218-225

[41] Khachatryan AG, Irman A, van Goor FA, Boller KJ. Femtosecond electron-bunch dynamics in laser wakefields and vacuum. Physical Review Accelerators and Beams. 2007; **10**:121301. DOI: 10.1103/PhysRevSTAB.10.121301

[42] Jackson JD. Classical Electrodynamics. 3rd ed. New York: John Wiley & Sons; 1999

[43] Michel P, Schroeder CB, Shadwick BA, Esarey E, Leemans WP. Radiative damping and electron beam dynamics in plasma-based accelerators. Physical Review E. 2006;**74**:026501. DOI: 10.1103/PhysRevE.74.026501

[44] Kirby N, Berry M, Blumenfeld I, Hogan MJ, Ischebeck R, Siemann R. Emittance growth from multiple coulomb scattering in a plasma wakefield accelerator. In: Proceedings of Particle Accelerator Conference 2007 (PAC2007); Albuquerque. New Mexico, USA: IEEE; 2007. pp. 3097-3099

[45] Deng A, Nakajima K, Liu J, Shen B, Zhang X, Yu Y, et al. Electron beam

dynamics and self-cooling up to PeV level due to betatron radiation in plasma-based accelerators. Physical Review Accelerators and Beams. 2012; **15**:081303. DOI: 10.1103/PhysRevSTAB.15.081303

[46] Ehrlich Y, Cohen C, Zigler A, Krall J, Sprangle P, Esarey E. Guiding of high intensity laser pulses in straight and curved plasma channel experiments. Physical Review Letters. 1996;**77**:4186-4189

[47] Reitsma AJW, Jaroszynski DA. Propagation of a weakly nonlinear laser pulse in a curved plasma channel. Physics of Plasmas. 2007;**14**:053104. DOI: 10.1063/1.2731816

[48] Luo J, Chen M, Wu WY, Weng SM, Sheng ZM, Schroeder CB, et al. Multistage coupling of laser-wakefield accelerators with curved plasma channels. Physical Review Letters. 2018; **120**:154801. DOI: 10.1103/PhysRevLett.120.154801

[49] Steinke S, Tilborg J, Benedetti C, Geddes CGR, Schroeder CB, Daniels J, et al. Multistage coupling of independent laser-plasma accelerators. Nature. 2016;**530**:190-193. DOI: 10.1038/nature16525

[50] Noble RJ. Beamstrahlung from colliding electron-positron beams with negligible disruption. Nuclear Instrumentation and Methods. 1987;**256**: 427-433

[51] Hollebeek R. Disruption limits for linear colliders. Nuclear Instruments & Methods. 1981;**184**:333-347

[52] Chen P, Yokoya K. Disruption effects from the interaction of round e^+e^- beams. Physical Review D. 1988;**38**: 987-1000

[53] Wheeler J, Mourou G. Tajima T. In: Chao AW, Chou W, editors. Review of Accelerator Science and Technology. Vol. 9. Singapore: World Scientific; 2016. pp. 151-163. DOI: 10.1142/S1793626816300073

Chapter 5

Theoretical Calculations of the Masses of the Elementary Fermions

Nathalie Olivi-Tran

Abstract

Our universe is three-dimensional and curved (with a positive curvature) and thus may be embedded in a four-dimensional Euclidean space with coordinates x, y, z, t where the fourth dimension time t is treated as a regular dimension. One can set in this spacetime a four-dimensional underlying array of small hypercubes of one Planck length edge. With this array all elementary particles can be classified following that they are two-, three-, or four-dimensional. The elementary wavefunctions of this underlying array are equal to $\sqrt{2}exp\,(ix_i)$ for $x_i = x, y, z$ or to $\sqrt{2}exp\,(it)$ for t. Hence, the masses of the fermions of the first family are equal to 2^n (in eV/c^2) where n is an integer. The other families of fermions are excited states of the fermions of the first family and thus have masses equal to $2^n.(p^2)/2$ where n and p are two integers. Theoretical and experimental masses fit within 10%.

Keywords: four-dimensional spacetime, masses of elementary fermions, theoretical masses, real space theory, Grand Unified Theory

1. Introduction

Since the beginning of the twentieth century, experimental particle physics has been making large progresses with the set up of accelerators and colliders.

The main locations of accelerators are presently: the Centre Europeen de Recherches Nucleaires (CERN) near Geneva (Switzerland and France). Equipments of the CERN are presently the Super Proton Synchrotron and the Large Hadron Collider (LHC), which is a protons collider. In Germany, the DESY (Deutsche Elektronen Synchrotron) main set up HERA is a collider between electrons or positrons and protons. In the USA, the Stanford Linear Accelerator Center (SLAC) main set up is PEP-II, which is a collider between electrons and positrons. Located also in the USA, the Fermi National Accelerator Laboratory (Fermilab) uses its main set up the Tevatron to collide protons and antiprotons. Finally, the Brookhaven National Laboratory (USA) uses the set up Relativistic Heavy Ion Collider to study collisions between heavy ions.

Up to now, the results obtained with colliders and accelerators fit the Standard Model, which predicts the existence of three families of elementary fermions and five different types of bosons. Although string theories [1] and supersymmetry [2] try to unify all different types of elementary particles, no experimental proof has

IntechOpen

been made of these theories. So, we present here a new theory that aims to unify elementary particles characteristics. The theoretical masses of the elementary particles are compared to the experimental masses.

This book chapter is a small review of the theoretical calculations of the masses of elementary particles in real space [3–5]. The theoretical masses that we calculated fit the experimental masses within less than 10% for almost all elementary fermions.

Our universe is three-dimensional and has a positive curvature. So our universe may be embedded in an Euclidean four-dimensional space. These four dimensions are x, y, z, t where t is time [6–8]. In this four-dimensional space, we classify the elementary particles following their geometry, that is, elementary particles may be four-, three-, or two-dimensional [5] (see Section 2). Let us notice that for a given particle, time and mass are linked: if the mass of a particle is zero, this particle has no temporal dimension.

If spacetime is composed of small hypercubes of one Planck length edge, there exist elementary wavefunctions that are equal to $\sqrt{2}\, exp\,(ix_i)$ if it corresponds to a space dimension or equal to $\sqrt{2}\, exp\,(it)$ if it corresponds to a time dimension (these elementary wave functions are obtained by calculating the eigenfunction of a particle in a one-dimensional box, that is, the edge of the underlying hypercubes). The masses of the electron, of the electron neutrino and of the quark up (first family of fermions) are equal to integer powers of 2 (in eV/c^2) [3]. We will show that the fermions of the second and third families are excited states of the fermions of the first family. Indeed, the masses of all elementary fermions follow the formula $2^n.(p^2)/2$ where n is an integer [3, 4] calculated for the electron, electron neutrino and quark up and p is another integer that corresponds to the excited states of the elementary wavefunctions (see Section 3).

2. Dimensions of elementary particles

All the theories that aim to understand the elementary particles treat time t as a special dimension. Thus, many physicists deal with $n + 1$ space dimensions in particle physics, where the +1 corresponds to the special temporal dimension, thus treated differently. As previously published [3–8], time may be seen as a function of space dimension, if our three-dimensional universe is embedded in a four-dimensional space (due to the positive curvature of our three-dimensional universe).

So, here we will present a simple hypothesis about the classification of elementary particles based on the fact that the space is four-dimensional and that time t is a dimension like x, y and z. Here, this book chapter is dedicated to our hypothesis. This classification is intuitive but next sections of this book chapter, which deal with the masses of the elementary fermions, use and thus demonstrate this classification.

Indeed, with simple arguments, it seems to lead to the Grand Unified Theory (GUT). Time is a function of the fourth dimension of this four-dimensional Euclidean space. If we apply this hypothesis to particle physics, we may say that elementary particles are four-dimensional, three-dimensional and two-dimensional. The coordinates (x, y, z, t) are not orthonormal. Indeed, time t evolves as $log\,(r)$ where r is the co-moving distance in cosmology [6]. Let us make the additional assumption that for each of these four dimensions there are functions like $exp\,(ir_j)$ with $(r_j = x, y, z, t)$ that vibrate (like in string theory). To find these elementary functions, one has to solve the one-dimensional problem of a particle in a square potential of edge length $\hbar$ (the Planck constant).

So, our reasoning is simply the description of how to distribute these functions in the four-dimensional space. In the following, the reasoning applies in real space.

A previous paper of mine (see [3, 9]) predicts that the Higgs potential in real space is a hypercubic box in our four-dimensional space. To obtain the first family of fermions from the Standard Model (i.e., quark up, electron, electron neutrino), one may say that see **Figure 1**:

- the electron is four-dimensional (t, x, y, z);
- the quark up is three-dimensional (t, x, y) or (t, x, z) or (t, y, z); and
- the electronic neutrino is two-dimensional (t, x) and x, y and z are equivalent. When this neutrino propagates, there are infinitesimal rotations between the characteristic coordinates (leading to flavor oscillations).

To obtain the masses of the remaining fermions (fermions of the second and third families), one has to add a second quantum number p (similar to the quantum number obtained for a particle in a square potential of dimensions $\hbar$—the Planck length). Thus, the remaining fermions of the Standard Model may be seen as excited states of the first fermion family.

Bosons may be classified with the same assumptions see **Figure 2**:

- the photon is two-dimensional (x, y) but has no temporal t coordinate—no mass (indeed with my Higgs potential [3, 9], time at square is proportional to the mass);
- the gluon is three-dimensional (x, y, z) and has no temporal dimension—no mass (during the strong interaction, one gluon interferes (positive interferences) with two quarks: x on x, y on y, etc.);
- the Z and W bosons are three-dimensional with mass (t, x, y); and
- the Higgs boson is four-dimensional (x, y, z, t).

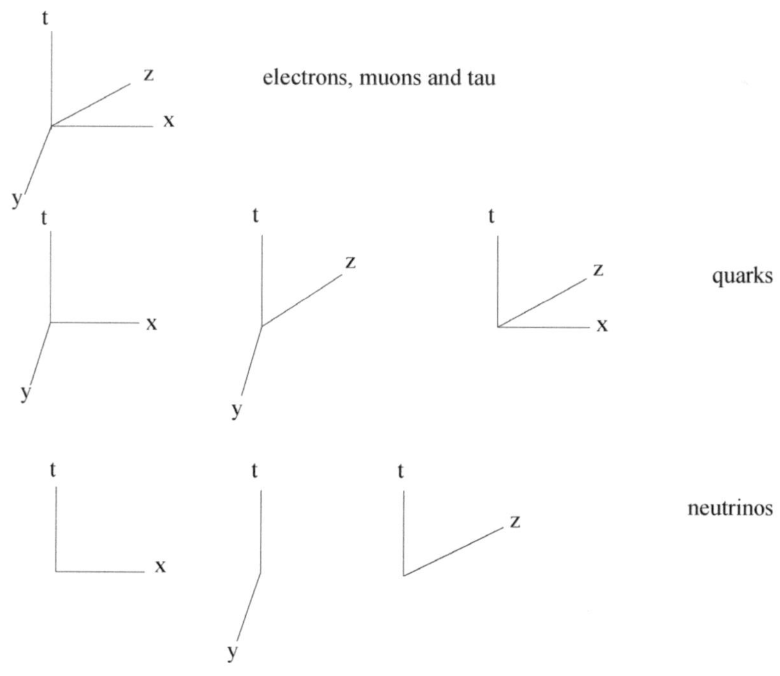

Figure 1.
Dimensions of elementary fermions.

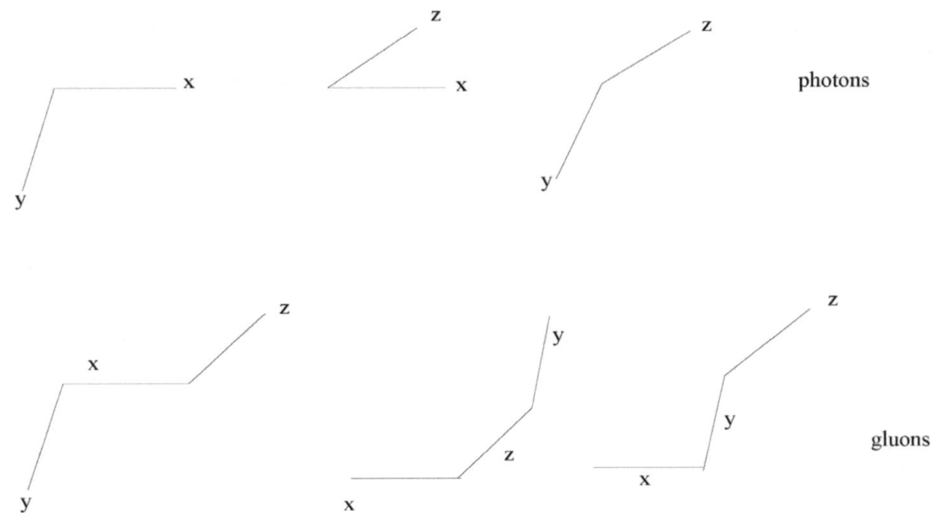

Figure 2.
Dimensions of elementary bosons.

In all these descriptions [5], the geometrical characteristics of elementary particles have been separated from their equation of propagation. With this hypothesis, we obtained a new geometrical classification of elementary particles. Presently, most of the calculations have been made using Feynman graphs, that is, in the space of functions, leading to symmetries that are not yet unified. So in our opinion, the symmetries in the Standard Model do not give the entire description of elementary particles.

In the following section, I will use the geometrical dimensions of the elementary particles to calculate the masses of elementary particles.

3. Masses of elementary fermions

In quantum mechanics, the wavefunction gives the most fundamental description of the behavior of a particle; the measurable properties of the particle (such as its position, momentum and energy) may all be derived from the wavefunction. The wavefunction $\psi(x,t)$ can be found by solving the Schrödinger equation for the system [10].

$$i\hbar\frac{\partial}{\partial t}\psi(x,t) = -\frac{\hbar^2}{2m}\frac{\partial^2}{\partial x^2}\psi(x,t) + V(x,t)\psi(x,t) \tag{1}$$

where $\hbar$ is the reduced Planck constant, m is the mass of the particle, i is the imaginary unit and t is time. The square potential $V(x,t)$ is equal to zero for $x < L$ and $x > 0$ and for $t < L$ and $t > 0$. We use Von-Karman boundary conditions. Moreover, the domain of definition of the function $\psi(x,t)$ is $[0,L]$ for x and also $[0,L]$ for t where L is the width of the potential $V(x,t)$.

The eigenfunctions of the Schrödinger equation may be written:

$$\psi(x,t) = \sqrt{\frac{2}{L}}exp\left(-ik_p x\right)exp\left(-i\omega t\right) \tag{2}$$

where $\sqrt{\frac{2}{L}}$ normalizes the eigenfunctions. To compute the energy levels of these eigenfunctions, we have:

$$k_p = \frac{p\pi}{L} \tag{3}$$

and

$$E_p = \hbar\omega_p = \frac{p^2\pi^2\hbar^2}{2mL^2}, \tag{4}$$

In order to simplify our calculations, we normalize all constants so that the eigenfunctions are equal to:

$$\psi(x,t) = \sqrt{2}exp\,(ix_j)\,exp\,(it) \tag{5}$$

To obtain the masses of all elementary fermions (elementary particles), one has to modify the quantum number p [4] (similar to the quantum number of a particle in a box). Thus, the remaining fermions of the Standard Model may be seen as excited states of the first fermion family.

Straightforwardly, we make the following hypotheses:

- spacetime has an underlying hypersquare array of edge length $\hbar$;
- elementary wave functions (in (x,y,z,t) space) are eigenfunctions of a particle in a square potential (reduced parameters) $\sqrt{2}exp\,(-ix)$ for space $\sqrt{2}exp\,(-it)$ for time; and
- the eigenvalues of the elementary wave functions are equal to $\frac{p^2}{2}$ (with p an integer number).

In the following subsection, I will use the preceding hypotheses to calculate theoretically the masses of the elementary fermions.

3.1 Masses of the electron, muon and tau

The Dirac equation may be written:

$$i\gamma^\mu\partial_\mu\psi - m\psi = 0 \tag{6}$$

with ψ the wavefunction, m the mass of the fermion and with the Dirac matrices:

$$\gamma^0 = \begin{pmatrix} I_2 & 0 \\ 0 & -I_2 \end{pmatrix}, \tag{7}$$

$$\gamma^1 = \begin{pmatrix} 0 & \sigma_x \\ -\sigma_x & 0 \end{pmatrix}, \tag{8}$$

$$\gamma^2 = \begin{pmatrix} 0 & \sigma_y \\ -\sigma_y & 0 \end{pmatrix}, \tag{9}$$

$$\gamma^3 = \begin{pmatrix} 0 & \sigma_z \\ -\sigma_z & 0 \end{pmatrix}, \tag{10}$$

where σ_ν are the Pauli matrices.

Using combinatorial analysis, we obtain Eq. (11) (using the fact that electrons are $4d$ [5] and that all space dimensions are equivalent).

$$\begin{pmatrix} \gamma_1 & 0 & 0 & 0 & 0 & 0 & 0 & 0 & 0 & 0 \\ 0 & \gamma_2 & 0 & 0 & 0 & 0 & 0 & 0 & 0 & 0 \\ 0 & 0 & \gamma_3 & 0 & 0 & 0 & 0 & 0 & 0 & 0 \\ 0 & 0 & 0 & \gamma_3 & 0 & 0 & 0 & 0 & 0 & 0 \\ 0 & 0 & 0 & 0 & \gamma_1 & 0 & 0 & 0 & 0 & 0 \\ 0 & 0 & 0 & 0 & 0 & \gamma_2 & 0 & 0 & 0 & 0 \\ 0 & 0 & 0 & 0 & 0 & 0 & \gamma_2 & 0 & 0 & 0 \\ 0 & 0 & 0 & 0 & 0 & 0 & 0 & \gamma_3 & 0 & 0 \\ 0 & 0 & 0 & 0 & 0 & 0 & 0 & 0 & \gamma_1 & 0 \\ 0 & 0 & 0 & 0 & 0 & 0 & 0 & 0 & 0 & \sigma_0 \end{pmatrix} \begin{pmatrix} \partial_1\Psi \\ \partial_2\Psi \\ \partial_3\Psi \\ \partial_1\Psi \\ \partial_2\Psi \\ \partial_3\Psi \\ \partial_1\Psi \\ \partial_2\Psi \\ \partial_3\Psi \\ \partial_0\Psi \end{pmatrix} = m. \begin{pmatrix} \Psi \\ \Psi \\ \Psi \\ \Psi \\ \Psi \\ \Psi \\ \Psi \\ \Psi \\ \Psi \\ \Psi \end{pmatrix} \tag{11}$$

There are three possibilities of arranging $\gamma_1, \gamma_2, \gamma_3$ (the Dirac matrices) over x, y and z (all space dimensions are equivalent) and one possibility to arrange σ_0 (temporal Pauli matrix: half of γ_0; because time does not go backward).

The large matrix M (see Eq. (11)) containing all combinations has a dimension $9X4 + 2 = 38$. We see that, with the coordinate vectors $\sqrt{2}exp\,(-it)$ and $\sqrt{2}exp\,(-ix)$ (eigenfunctions of a particle in a square potential), we have to multiply the modified Dirac equation by the Jacobian corresponding to these new coordinates. This Jacobian is equal to $\sqrt{2}^{38}$ where 38 is the dimension of the large matrix [3]. We multiply the mass of the first particle of this family by the eigenvalues of the eigenfunctions (of the particle).

We decompose the eigenvalues into prime numbers [4]. The number of eigenvalues for the ground state (electron) is 38 (the dimension of the large matrix M). For the other particles, we take into account the spinor $(1, 0)^T$ corresponding to the σ_0 Pauli matrix. So except for the electron, there are 37 eigenvalues for each particle [4].

- The mass of the electron is equal to $\sqrt{2}^{38} = 2^{19}eV/c^2 = 2^{19}.\left(\frac{1^2}{2}\right)^{19}.\left(\frac{2^2}{2}\right)^{19} = 0.524MeV/c^2 \approx 0.511\,MeV/c^2$.
- The mass of the muon is equal to $2^{19}.20^2/2 = 2^{19}.\frac{2^2}{2}.\frac{2^2}{2}.\frac{2^2}{2}.\frac{2^2}{2}.\frac{5^2}{2}.\left(\frac{1^2}{2}\right)^{16}.\left(\frac{2^2}{2}\right)^{16} = 104.8MeV/c^2 \approx 105.6\,MeV/c^2$.
- The mass of the tau is equal to $2^{19}.82^2/2 = 2^{19}.\frac{41^2}{2}.\frac{2^2}{2}.\frac{2^2}{2}.\left(\frac{1^2}{2}\right)^{17}.\left(\frac{2^2}{2}\right)^{17} = 1.76GeV/c^2 \approx 1.78\,GeV/c^2$.

The values in italic are the experimental masses [11].

We see that for the tau particle, one of the eigenvalues $\left(\frac{41^2}{2}\right)$ is much larger than the others. This may explain the short lifetime of this particle.

particle	Theoretical mass (eV/c^2)	Experimental mass (eV/c^2) [11]
electron	$2^{19} = 0.524MeV/c^2$	$0.511MeV/c^2$
muon	$2^{19}.20^2/2 = 104.8MeV/c^2$	$105.6MeV/c^2$
tau	$2^{19}.82^2/2 = 1.76GeV/c^2$	$1.78GeV/c^2$

Table 1.
Theoretical and experimental masses of the electron, muon and tau.

The masses (theoretical and experimental) of the electron, muon and tau are summarized in **Table 1.**

3.2 Masses of the quarks

For quarks, we have

$$i\gamma^{\mu}\partial_{\mu}\psi = m\psi \tag{12}$$

The Dirac matrices are representative of infinitesimal rotations within the wavefunction of a given elementary particle.

Using combinatorial analysis, we obtain Eq. (13) (using the fact that quarks are $3d$ [5] and that all space dimensions are equivalent). There are three possibilities for arranging $\gamma_1, \gamma_2, \gamma_3$ (the Dirac matrices) over x, y and z (all space dimensions are equivalent). There is one possibility to arrange σ_0 (temporal Pauli matrix; half of γ_0, because time does not go backward) for each combination of spatial Dirac matrices (x, y; x, z; and y, z). We have to take into account that the quarks are three-dimensional. So, the matrix M containing all combinations has a dimension equal to $9X4 + 3X2 = 42$.

$$\begin{pmatrix} \gamma_1 & 0 & 0 & 0 & 0 & 0 & 0 & 0 & 0 & 0 & 0 & 0 \\ 0 & \gamma_2 & 0 & 0 & 0 & 0 & 0 & 0 & 0 & 0 & 0 & 0 \\ 0 & 0 & \gamma_3 & 0 & 0 & 0 & 0 & 0 & 0 & 0 & 0 & 0 \\ 0 & 0 & 0 & \sigma_0 & 0 & 0 & 0 & 0 & 0 & 0 & 0 & 0 \\ 0 & 0 & 0 & 0 & \gamma_2 & 0 & 0 & 0 & 0 & 0 & 0 & 0 \\ 0 & 0 & 0 & 0 & 0 & \gamma_3 & 0 & 0 & 0 & 0 & 0 & 0 \\ 0 & 0 & 0 & 0 & 0 & 0 & \gamma_1 & 0 & 0 & 0 & 0 & 0 \\ 0 & 0 & 0 & 0 & 0 & 0 & 0 & \sigma_0 & 0 & 0 & 0 & 0 \\ 0 & 0 & 0 & 0 & 0 & 0 & 0 & 0 & \gamma_3 & 0 & 0 & 0 \\ 0 & 0 & 0 & 0 & 0 & 0 & 0 & 0 & 0 & \gamma_1 & 0 & 0 \\ 0 & 0 & 0 & 0 & 0 & 0 & 0 & 0 & 0 & 0 & \gamma_2 & 0 \\ 0 & 0 & 0 & 0 & 0 & 0 & 0 & 0 & 0 & 0 & 0 & \sigma_0 \end{pmatrix} \begin{pmatrix} \partial_1\Psi \\ \partial_2\Psi \\ \partial_3\Psi \\ \partial_0\Psi \\ \partial_1\Psi \\ \partial_2\Psi \\ \partial_3\Psi \\ \partial_0\Psi \\ \partial_1\Psi \\ \partial_2\Psi \\ \partial_3\Psi \\ \partial_0\Psi \end{pmatrix} = m \begin{pmatrix} \Psi \\ \Psi \\ \Psi \\ \Psi \\ \Psi \\ \Psi \\ \Psi \\ \Psi \\ \Psi \\ \Psi \\ \Psi \\ \Psi \end{pmatrix} \tag{13}$$

We see that, with the coordinate vectors $\sqrt{2}exp(-it)$ and $\sqrt{2}exp(-ix)$ (eigenfunctions of the underlying hypersquare array), we have to multiply the modified Dirac equation by the Jacobian corresponding to these new coordinates. This Jacobian is equal to $\sqrt{2}^{42}$ where 42 is the dimension of the matrix [3]. We multiply the mass of the first particle of the quarks family by the eigenvalues of the eigenfunctions (of the particle). We decompose the eigenvalues into prime numbers [4]. The number of eigenvalues for the ground state (quark up) is 42 (the dimension of the large matrix M, see Eq. (13)). For the other quarks, we take into account the spinor $(1,0)^T$ corresponding to the three σ_0 Pauli matrices. So except for the quark up, there are 39 eigenvalues for each quark [4].

- The quark up has a mass equal to $\sqrt{2}^{42} = 2^{21}eV/c^2 = 2^{21}.\left(\frac{1^2}{2}\right)^{21}.\left(\frac{2^2}{2}\right)^{21} = 2.09MeV/c^2 \approx 2.2\ MeV/c^2$.
- The quark down has a mass equal to $2^{21}.\frac{2^2}{2} = 2^{21}.\frac{2^2}{2}.\left(\frac{1^2}{2}\right)^{19}.\left(\frac{2^2}{2}\right)^{19} = 4.19MeV/c^2 \approx 4.7MeV/c^2$.
- The quark strange has a mass equal to $2^{21}.\frac{9^2}{2} = 2^{21}.\frac{3^2}{2}.\frac{3^2}{2}.\frac{2^2}{2}.\left(\frac{1^2}{2}\right)^{18}.\left(\frac{2^2}{2}\right)^{18} = 84.9MeV/c^2 \approx 96\ MeV/c^2$.
- The quark charm has a mass equal to $2^{21}.\frac{36^2}{2} = 2^{21}.\frac{3^2}{2}.\frac{3^2}{2}.\frac{2^2}{2}.\frac{2^2}{2}.\frac{2^2}{2}.\frac{2^2}{2}.\frac{2^2}{2}\left(\frac{1^2}{2}\right)^{16}.\left(\frac{2^2}{2}\right)^{16} = 1.35GeV/c^2 \approx 1.27\ GeV/c^2$.
- The quark bottom has a mass equal to $2^{21}.\frac{63^2}{2} = 2^{21}.\frac{3^2}{2}.\frac{3^2}{2}.\frac{2^2}{2}.\frac{2^2}{2}.\frac{7^2}{2}.\left(\frac{1^2}{2}\right)^{17}.\left(\frac{2^2}{2}\right)^{17} = 4.16GeV/c^2 \approx 4.18\ GeV/c^2$.
- The quark top has a mass equal to $2^{21}.\frac{405^2}{2} = 2^{21}.\frac{3^2}{2}.\frac{3^2}{2}.\frac{3^2}{2}.\frac{3^2}{2}.\frac{5^2}{2}.\frac{2^2}{2}.\frac{2^2}{2}.\frac{2^2}{2}.\frac{2^2}{2}.\left(\frac{1^2}{2}\right)^{15}.\left(\frac{2^2}{2}\right)^{15} = 171.9GeV/c^2 \approx 173\ GeV/c^2$.

The values in italic are the experimental masses [11].

The theoretical and experimental masses of the quarks family are summarized in **Table 2**.

quark	Theoretical mass (eV/c^2)	Experimental mass (eV/c^2) [11]
up	$2^{21} = 2.09MeV/c^2$	$2.2MeV/c^2$
down	$2^{21}.2^2/2 = 4.19MeV/c^2$	$4.7MeV/c^2$
strange	$2^{21}.9^2/2 = 84.9MeV/c^2$	$96MeV/c^2$
charm	$2^{21}.36^2/2 = 1.35GeV/c^2$	$1.27GeV/c^2$
bottom	$2^{21}.63^2/2 = 4.16GeV/c^2$	$4.18GeV/c^2$
top	$2^{21}.405^2/2 = 171.9GeV/c^2$	$173GeV/c^2$

Table 2.
Theoretical and experimental masses of the quarks family.

particle	Theoretical mass	Experimental mass: upper limit [11]
electron neutrino	$2eVc^2$	$2.5eVc^2$
muon neutrino	$412^2 = 169.7keVc^2$	$170keVc^2$
tau neutrino	$3937^2 = 15.4MeVc^2$	$18MeVc^2$

Table 3.
Theoretical masses of the neutrinos and upper limits of experimental masses.

3.3 Masses of the neutrinos

Up to now, there is no theoretical propagation equation for the neutrinos.

If we use the eigenvalues of the elementary wave functions like for quarks and electrons, muons and taus, we may write [4]:

- the mass of the electron neutrino is equal to $2eV/c^2$;
- the mass of the muon neutrino is equal to $2.\frac{412^2}{2} = 412^2 eV/c^2 = 169keV/c^2$; and
- the mass of the tau neutrino is equal to $2.\frac{3937^2}{2} = 3937^2 eV/c^2 = 15.4MeV/c^2$.

Hence, we found theoretical values of the masses of the neutrinos, which are in good agreement with the experimental masses (**Table 3**).

4. Conclusion

In this chapter, the calculations of the masses of all the known elementary fermions are made in real space. At the beginning of this book chapter (Section 2), I presented a classification of elementary particles over all space and temporal dimensions. Using this geometrical classification (which is intuitive), we found the theoretical values of masses for all the elementary fermions (electrons, muons and taus; all quarks and all neutrinos). The theoretical masses are in good agreement with the experimental masses (the differences between theoretical and experimental masses are less than 10% except for the quarks down and strange). To conclude, our theory unifies all elementary fermions: we use the same approach to all these fermions (geometry and the underlying hypersquare array of spacetime). In the future, there is a possibility to analyze the symmetries of these particles and compare them to the symmetries of the Standard Model.

Author details

Nathalie Olivi-Tran
Laboratoire Charles Coulomb, Universite de Montpellier, CNRS, Montpellier, France

*Address all correspondence to: nathalie.olivi-tran@umontpellier.fr

References

[1] Green MB, Schwarz JH, Witten E. Superstring Theory. Cambridge: Cambridge University Press; 1987. ISBN: 0521323843

[2] Peskin MS. Supersymmetry in Elementary Particle Physics arXiv: 0801.1928

[3] Olivi-Tran N. The masses of the first family of fermions and of the Higgs boson are equal to integer powers of 2. Nuclear and Particle Physics Proceedings. 2015;**258–259**:272-275. DOI: 10.1016/j.nuclphysbps.2015.01.057

[4] Olivi-Tran N. Theoretical approach to the masses of elementary fermions. Nuclear and Particle Physics Proceedings. 2020;**309-311C**:73-76

[5] Olivi-Tran N, Gottiniaux N. A classification of elementary particles in $d = 4$ following a simple geometrical hypothesis in real space. Advanced Studies in Theoretical Physics. 2013; **7**(18):853-857. DOI: 10.12988/astp.2013.3778

[6] Olivi-Tran N, Gauthier PM. The FLRW cosmological model revisited: Relation on the local time with the local curvature and consequences on the Heisenberg uncertainty principle. Advanced Studies in Theoretical Physics. 2008;**2**(6):267-270. DOI: 10.12988/astp

[7] Olivi-Tran N. What if our three dimensional curved universe was embedded in four dimensional space? Consequences on the EPR paradox. Advanced Studies in Theoretical Physics. 2009;**3**(12):489-492. DOI: 10.12988/astp

[8] Olivi-Tran N. Dimensional analysis of Einstein's fields equations. Advanced Studies in Theoretical Physics. 2009; **3**(1):9-12. DOI: 10.12988/astp

[9] Olivi-Tran N. Is it the Higgs scalar field? Advanced Studies in Theoretical Physics. 2010;**4**(13):633-636. DOI: 10.12988/astp

[10] https://en.wikipedia.org/wiki/Particle_in_a_box

[11] Patrignani C et al. (Particle Data Group). Review of Particle Physics. Chinese Physics C. 2016;**40**:100001. DOI: 10.1088/1674-1137/40/10/100001

Chapter 6

Review of Quantum Chromodynamics (QCD)

Leonard S. Kisslinger

Abstract

This is a review of the elementary particles, quantum chromodynamics (QCD), and strong interactions in QCD theory via gluon exchange between quarks-antiquarks-producing mesons. Some mesons consist of an active gluon in addition to a quark-antiquark. They are called hybrid mesons. We also review the possible detection of the quark-gluon plasma, the consistuent of the universe until about 10^{-4} s after the Big Bang, via relativistic heavy ion collisions (RHIC) producing heavy quark hybrid mesons.

Keywords: quantum chromodynamics, elementay particles, quark-gluon plasma

1. Introduction

In this review of quantum chromodynamics (QCD), basic QCD particles and forces, the theory of hybrid mesons, the cosmological quantum chromodynamics phase transition (QCDPT), the quark-gluon plasma (QGP), theoretial studies of jet quenching due to the formation of the QGP, and the possible detection of the QGP via the production of mixed hybrid meson states produced by RHIC are reviewed.

First we review elementary particles, fermions and bosons, and standard QCD theory with the concept of color for the basic fermions (quarks) and bosons (gluons).

In the next section, the structure of standard mesons in terms of quarks and antiquarks is discussed.

Next we review the theory predicting that some mesons consisting of heavy charmonium quarks ($\Psi(nS)$, n = 1, 2) and bottomonium quarks ($\Upsilon(mS)$, m = 1, 2, 3, 4) are partially hybrids. Experiments test the theory.

The interactions in quantum chromodynamics are strong, so perturbation theory does not work. Therefore, Feynman diagrams used for quantum electrodynamics cannot be used for quantum chromodynamics [1].

One nonperturbative QCD method involves lattice gauge theory. The articles "Twenty-first Century Lattice Gauge Theory: Results from the QCD Lagrangian" by Kronfelld [2] and "Lattice gauge theory in the microcanonical ensemble" by Calloway and Rahman [3] give a detailed description of how lattice gauge theory can calculate QCD interactions using computers.

Another nonperturbative theory, which is used in our review of mixed hybrid heavy quark mesons, is the method of QCD sum rules. This method, developed by Shifman et al. [4], does not require large computers.

IntechOpen

It was the method of QCD sum rules that showed [5] that the $\Psi(2S)$ charmonium quark meson and the $\Upsilon(3S)$ bottomonium quark mesons are mixed hybrid states, while all the other $\Psi(nS)$ and $\Upsilon(mS)$ are standard charmonium and bottomonium meson states.

In our final section, we review the early universe QCDPT, with the production of the QGP.

In our second subsection, theoretial studies of jet quenching due to the formation of the QGP and experiments with Pb-Pb collisions verifying the theory are discussed.

In our final section, the possible production of the QGP via relativistic heavy ion collisions (RHIC) with the possible detection of the QGP by the production of mixed heavy quark hybrid mesons is reviewed. That is, we consider the collision of gold (Au) atomic nuclei with the energy of the Au nuclei large enough that after the Au-Au collision the temperature of the overlapping material is $T \geq T_c^{QCDPT} \simeq 150$ MeV, where T_c^{QCDPT} is the critical temperature for the QCDPT. That matter with $T \geq T_c^{QCDPT}$ is QGP is discussed.

The production of the mixed hybrid states $\Psi(2S)$ and $\Upsilon(3S)$ via Au-Au collisions could detect the production of the QGP in the overlapping material. We conclude that the detection of $\Psi(2S)$ and $\Upsilon(3S)$ produced via RHIC could be a test for the creation of the early universe QGP.

2. Elementary particles and basic forces

Among the elementary particles—fermions and bosons—fermions have quantum spin = 1/2.

The elementary femions are leptons and quarks. There are three generations of leptons: electron, muon, and tau, with electric charge −1, and their neutrinos with no electric charge. There are three generations of quarks: (u, d); (c, s); and (t, b). The (u, c, t) quarks have electric charge 2/3 while the (d, s, b) quarks have electric charge −1/3.

First Generation	$\begin{pmatrix} e^- \\ \nu^e \end{pmatrix}$	$\begin{pmatrix} u \\ d \end{pmatrix}$
Second Generation	$\begin{pmatrix} \mu^- \\ \nu^\mu \end{pmatrix}$	$\begin{pmatrix} c \\ s \end{pmatrix}$
Third Generation	$\begin{pmatrix} \tau^- \\ \nu^\tau \end{pmatrix}$	$\begin{pmatrix} t \\ b \end{pmatrix}$

Bosons have quantum spin = 1: photon, quantum of the electromagnetic field; gluon, quantum of the strong field; and W and Z, weak field quanta, which we do not need.

2.1 Quantum chromodynamics (QCD): strong interaction field theory

Strong interactions are produced by quarks exchanging gluons, as illustrated in the figure below.

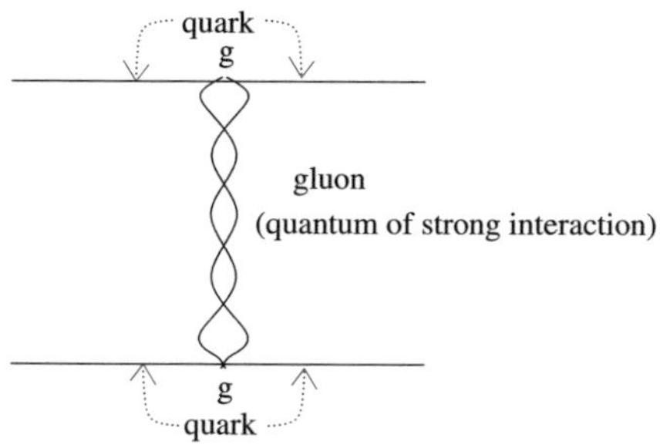

QCD (Quantum Chromodynamics): quark force via gluon exchange

STRONG FORCE

$g^2 \sim 1 \sim 100 \times e^2$

Nonperturbative. Diagrams do not converge– single diagram no good

Color: Quarks have three colors.

Color is to the strong interaction as electric charge is to the electromagnetic interaction.

A quark and antiquark can form a gluon, which has color 8.

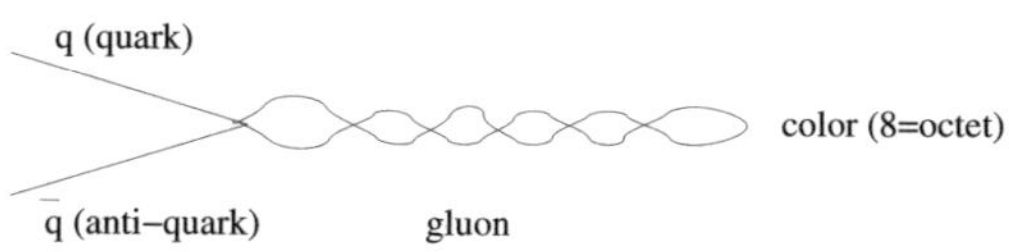

Note that particles with color, like gluon and quarks, cannot move freely in space. Particles that can move freely are baryons, like the proton and neutron, and mesons, which have no total color.

Note that in the figure above, the quark and antiquark have a different color, which is why the gluon has color = 8.

Antiparticles: All fermions have antiparticles. The antiparticle of the electron e^- is the positron e^+, and a photon can create a e^-e^+ pair. Similarly, each quark q has an antiquark $\overline{q}$.

Standard mesons consist of a quark and an antiquark. A meson that is important for today's discussion is the $J/\Psi(1S)$.

$$|J/\Psi(1S)> \; = |c\bar{c}(1S)> . \tag{1}$$

As we discuss below, the first excited $c\bar{c}$ state, the $\Psi(2S)$ has a hybrid component with an active gluon: $|c\bar{c}g(2S)>$.

3. QCD sum rules and mixed heavy quark hybrid meson states

The starting point of the method of QCD sum rules [4] is the correlator

$$\Pi^A(x) = \langle |T[J_A(x)J_A(0)]| \rangle, \tag{2}$$

with $|\rangle$ the vacuum state and the current $J_A(x)$ creates the states with quantum numbers A. For the charmonium states, J_c is

$$J_c = f J_{c\bar{c}} + \sqrt{1-f^2} J_{c\bar{c}g}, \tag{3}$$

where $J_{c\bar{c}}$ creates a normal charmonium state and $J_{c\bar{c}g}$ creates a hybrid state with an active gluon. It was shown that $f \simeq -\sqrt{2}$ for the $\Psi(2S)$ and $\Upsilon(3S)$ and $f \simeq 1.0$ for the other charmonium and bottomonium states [5]. Therefore,

$$\begin{aligned} |J/\Psi(1S)> &\simeq |c\bar{c}(1S)> \\ |\Psi(2S)> &\simeq -\sqrt{2}|c\bar{c}(1S)> + \sqrt{2}|c\bar{c}g(2S)> \\ |\Upsilon(3S)> &\simeq -\sqrt{2}|b\bar{b}(3S)> + \sqrt{2}|b\bar{b}g(3S)>, \end{aligned} \tag{4}$$

That is $J/\Psi(1S)$ is a standard charmonium meson and all of the $\Upsilon(nS)$ mesons are standard bottominium mesons except for $n = 3$.

3.1 Experimental verification that the $\Psi(2S)$ and $\Upsilon(3S)$ are mixed hybrid heavy quark mesons

Ψ and Υ production by Cu-Cu collisions for E = 200 GeV [6] are shown in **Figures 1–4**. The dashed curves are for the standard model.

Tests of the mixed hybrid theory for $\Psi(2S)$ and $\Upsilon(3S)$ states using ratios of cross sections for Cu-Cu collisions at E = 200 GeV: Since the absolute magnitude of dσ/dy for production of $\Psi(2S)$ states via Cu-Cu collisions is not certain, due to uncertainty in the normalization of the states, the tests of the theory [6] were carried out using ratios of cross sections, which can be compared to experiments.

From **Figures 1** and **2**, the ratios of $\Psi(2S)$ to $J/\Psi(1S)$ for the standard model (st) and the mixed hybrid theory (hy) for A-A (including Cu-Cu) collisions are

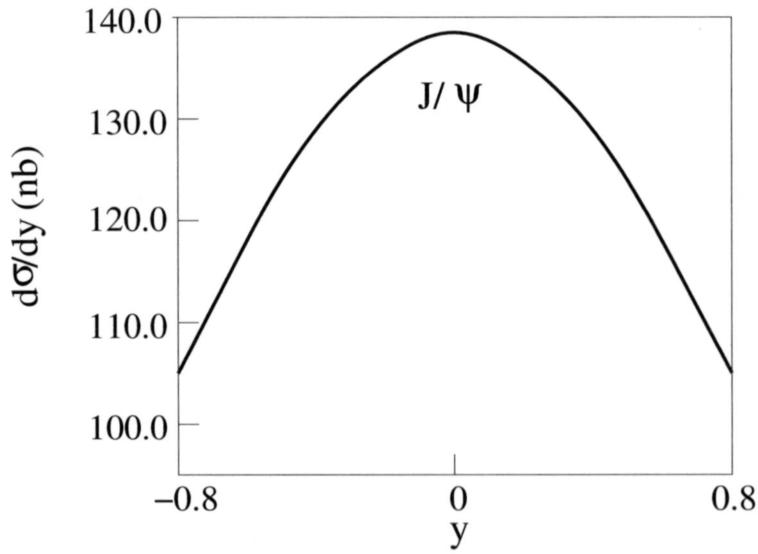

Figure 1.
dσ/dy for E = 200 GeV Cu-Cu collisions producing J/Ψ.

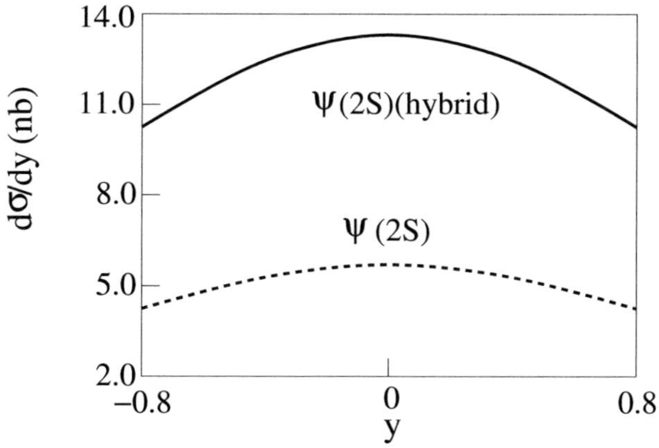

Figure 2.
dσ/dy for E = 200 GeV Cu-Cu collisions producing Ψ(2S).

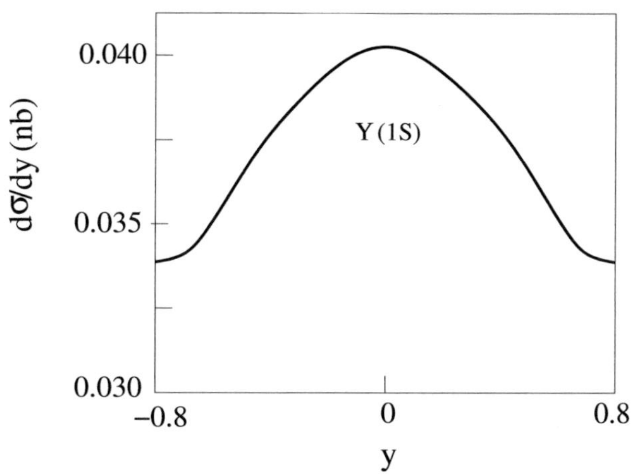

Figure 3.
dσ/dy for E = 200 GeV Cu-Cu collisions producing Υ(1S).

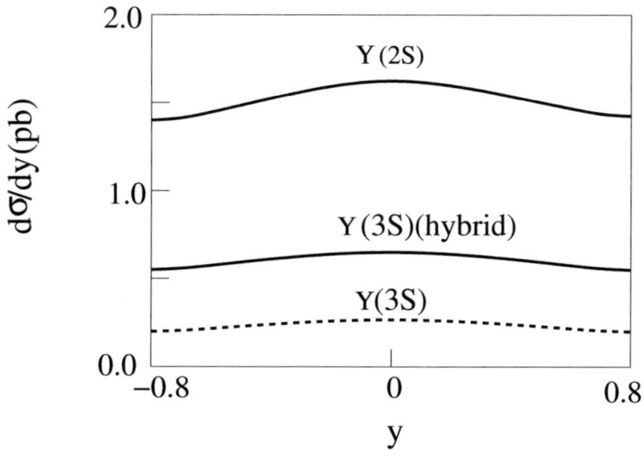

Figure 4.
dσ/dy for Cu-Cu collisions producing Υ(2S), Υ(3S).

$$\sigma(\Psi(2S))/\sigma(J/\Psi(1S))|_{st-A-A} \simeq 0.27$$
$$\sigma(\Psi(2S))/\sigma(J/\Psi(1S))|_{hy-A-A} \simeq 0.52 \pm 0.05, \quad (5)$$

while the experimental result [7] is

$$\sigma(\Psi(2S))/\sigma(J/\Psi(1S) \simeq 0.59, \quad (6)$$

which shows that the mixed hybrid theory for the $\Psi(2S)$ state is consistent with experiment, while the standard $|c\bar{c}(2S)>$ is not. From studies of heavy quark state production in p-p collisions, theoretical results for the nature of $\Upsilon(3S)$ state found [8]

$$\sigma(\Upsilon(3S))/\sigma(\Upsilon(1S)) \simeq 0.04 \;\; standard$$
$$\sigma(\Upsilon(3S))/\sigma(\Upsilon(1S)) \simeq 0.147 - 0.22 \;\; hybrid \quad (7)$$

compared to the experimental result of about 0.12–0.16 [9]. Therefore, the $\Upsilon(3S)$ state, as well as the $\Psi(2S)$ state, has been shown to be a heavy quark mixed hybrid state.

4. The cosmological QCDPT and possible detection of the QGP

In this section we first review the temperature of the universe based on the time when the cosmolical quantum chromodynamic phase transition (QCDPT) occured. At that time, the universe consisted of the quark-gluon plasma (QGP). Then we discuss the creation of the QGP via RHIC Au-Au collisions with the possible detection of the QGP.

4.1 The quantum chromodynamic phase transition (QCDPT)

From astrophysical studies, it is known that the QCDPT occured at a time $t \simeq 10^{-5} - 10^{-4}$ s; one can estimate the temperature of the universe at time T(t) using the equations from Einstein's General Theory of Relativity.

A simpler form of Einstein's equations are Friedman's equations. From Friedman's equations, one can find an approximation to T(t) [10].

$$T(t) \simeq \frac{1 \; MeV}{\sqrt{t(\text{in s})}}. \quad (8)$$

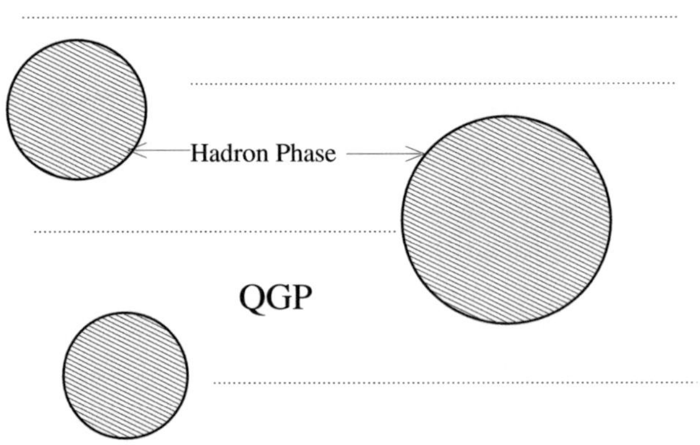

Figure 5.
Hadron phase forming within the QGP during the QCDPT.

From this, one finds, using $t \simeq 5 \times 10^{-5}$ s, at the time of the QCDPT the critical temperature for the QCDPT $T_c^{QCDPT} \simeq 150$ MeV. During the time that $T = T_c^{QCDPT}$, bubbles of our universe nucleated within the QGP, as shown in **Figure 5**.

4.2 Theoretical studies and predictions for the detection of the QGP

Theoretial studies of jet quenching due to the formation of the QGP were carried out.

Theoretial studies of jet quenching due to the formation of the QGP in high-energy Pb-Pb collisions [11] help motivate experimental studies. The theoretical equation predicting jet quenching [11] is

$$\Delta_{QGP} \simeq 1 - e^{-\frac{\Theta_{jet}}{\Theta_c}}, \tag{9}$$

where Δ_{QGP} is the magnitude of the jet quenching and Θ_{jet} and Θ_c are the aperature of the jet and emitted gluons, respectively.

Motivated in part by the theoretical study, the CMS collaboration carried out a study [12] of jet quencing via jet+Z boson correlations in Pb-Pb collisions. The results are shown in **Figure 6**.

As can be seen from the figure and Eq. (9), the theoretical prediction of jet quenching due to the QGP has been verified by experiments.

4.3 Creation and detection of the QGP via RHIC

A main goal of the study of heavy quark state production in relativistic heavy ion collisions (RHIC) is the detection of the quark-gluon plasma [13]. The energy of the atomic nuclei must be large enough so just after the nuclei collide, the temperature is that of the universe about 10^{-5} s after the Big Bang, when the universe was too hot for protons or neutrons and consisted of quarks and gluons (the constituents of proton and nucleons)—the quark-gluon plasma (QGP).

As **Figure 7** illustrates for Au-Au collisions with sufficient energy that the temperature $T(t) \geq 150$ MeV where the Au-Au nuclei merge, the emission of mixed

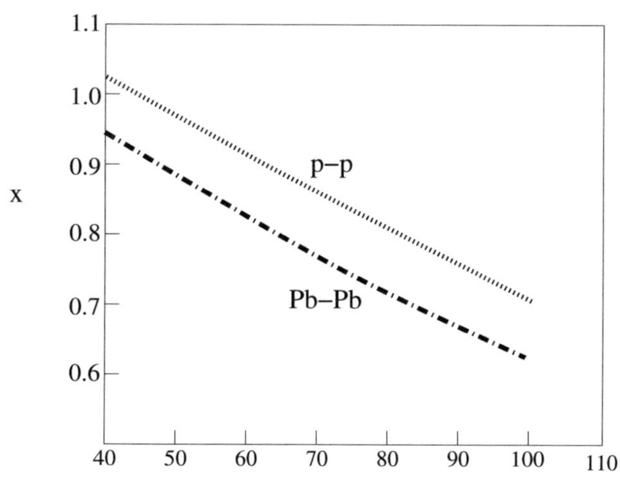

Figure 6.
x = (jet/Z) momentum vs. p^z = the momentum of the Z boson for p-p and Pb-Pb collisions.

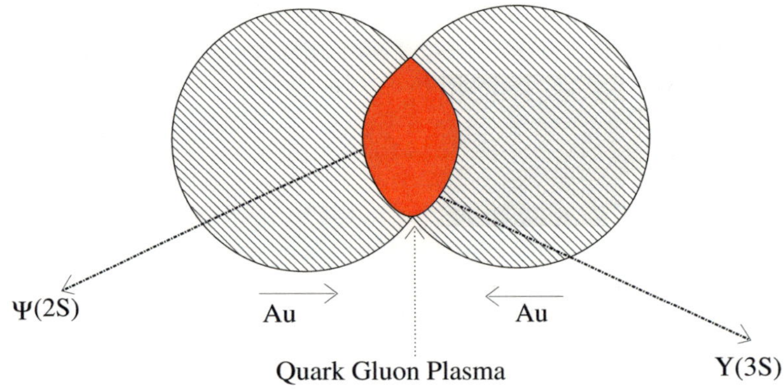

Figure 7.
Au-Au collisions producing $\Psi(2S)$ and $\Upsilon(3S)$ from the QGP.

hybrid mesons, the $\Psi(2S)$ and $\Upsilon(3S)$ as discussed above, with active gluons, could be a signal of the formation of the QGP.

5. Conclusions

The results from comparison of the production of $\Psi(2S)$ and $\Upsilon(3S)$ differential cross sections with experiment confirm the theoretical prediction that the $\Psi(2S)$ and $\Upsilon(3S)$ states are mixed hybrid states.

There are tests of the creation of the QGP from jet quenching via experiments using Z-jet correlations, as well as other tests suggested by theoretical studies.

There are also possible tests of the creation and detection of the QGP by the production of $\Psi(2S)$ and $\Upsilon(3S)$ via relativistic heavy ion collisions as shown in **Figure 7**.

From this, one can conclude that the production of these states via RHIC with sufficient energy that part of the matter during the collision has reached a temperature $T \geq T_c^{QCDPT} \simeq 150$ MeV, the temperature when the universe was a dense plasma (the QGP), can be a test of the creation of the quark-gluon plasma.

This would be an important result for particle theory as well as astrophysics.

Acknowledgements

Leonard S. Kisslinger acknowledges support in part as a visitor at the Los Alamos National Laboratory,

Author details

Leonard S. Kisslinger
Department of Physics, Carnegie Mellon University, Pittsburgh, PA, USA

*Address all correspondence to: kissling@andrew.cmu.edu

References

[1] Cheng T-P, Li L-F. Gauge Theory of Elementary Particle Physics. New York: Oxford University Press; 1985

[2] Kronfeld AS. Annual Review of Nuclear and Particle Science. 2012;**62**: 265

[3] Callaway DJE, Rahman A. Physical Review D. 1983;**28**:1506

[4] Shifman MA, Vainstein AI, Zakharov VI. Nuclear Physics B. 1979; **147**:385, 448

[5] Kisslinger LS. Physical Review D. 2009;**79**:114026

[6] Kisslinger LS, Liu MX, McGaughey P. Physical Review C. 2014; **89**:024914

[7] Adare A et al. Physical Review D. 2012;**85**:092004

[8] Kisslinger LS, Liu MX, McGaughey P. Physical Review D. 2011; **84**:114020

[9] Moreno G et al. Physical Review D. 1991;**43**:2815

[10] Kolb EW, Turner MS. The Early Universe. United States of America: Westview Press; 1990

[11] Casaiderry-Solana et al. Physical Letters B. 2013;**725**:357

[12] Sirunyan AM et al. Physical Review Letters. 2017;**119**:082301

[13] Kisslinger LS, Das D. International Journal of Modern Physics A. 2016;**31**: 1630010